Core Skills in Science

Photos courtesy of:

p. 44: © Stubblefield Photography; p. 45: © Darwin Online; p. 74: © CreativeNature R. Zwerver; p. 107: © Nerthuz; p. 115: © Denisa V; p. 129: © www.sandatlas.org; p. 131: © MarcelC; p. 135: © LackyVis; p. 141: © Atelopus; p. 148: © JamDay

Pre-HSE Core Skills in Science
ISBN 978-1-56420-879-8

New Readers Press
ProLiteracy's Publishing Division
104 Marcellus Street, Syracuse, New York 13204
www.newreaderspress.com

Printed in the United States of America
10 9 8 7 6 5 4 3

Proceeds from the sale of New Readers Press materials support professional development, training, and technical assistance programs of ProLiteracy that benefit local literacy programs in the U.S. and around the globe.

Developer: QuaraCORE
Editorial Director: Terrie Lipke
Cover Design: Carolyn Wallace
Technology Specialist: Maryellen Casey

CONTENTS

CONTENTS

INTRODUCTION

Welcome to *Core Skills in Science,* an important resource in helping you build a solid foundation of science skills as you gear up to start preparing for the GED®, TASC, or HiSET® high school equivalency Science test.

How to Use This Book

Pretest

The first step in using *Core Skills in Science* is to take the Pretest, which begins on the next page. This test will show which skills you already have and which areas you need to practice. After taking the Pretest and checking your answers, use the chart on page 12 to find the lessons that will help you study the skills you need to improve.

Science Skills Lessons

The book is organized into four units, each containing brief lessons that focus on a different area of science skills and knowledge. Read each of the lessons, which also include tips to help you understand skills and concepts:

- Real-World Connections describe how science concepts connect to our everyday lives.
- Helpful Hints offer suggestions for understanding key terms and reading charts and illustrations.

Important vocabulary terms are listed under Key Terms on the first page of each lesson. These words appear in **boldface** when they are first used in each lesson. You can use the Glossary at the end of the book to look up definitions of key terms.

Each lesson ends with a brief Lesson Review with questions to test your knowledge of the content that was covered. In the Answer Key, you will find explanations to help your understanding with many Lesson Review questions.

Each unit concludes with a Unit Practice Test that covers all the content in the unit's lessons.

Posttest

After completing all the units, you can test what you know by taking the Posttest, beginning on page 140. This test will help you check your understanding of all the skills in the book. It will also help determine if you are ready to move on to high school equivalency test preparation.

SCIENCE PRETEST

Answer the following questions to test your knowledge of science content and skills.

Use the following information and data table to answer questions 1–6.

Three equal amounts of different soil are placed in separate funnels. The types of soil are sandy soil, clay, and loamy. 100 mL of water is poured through each funnel and collected in a beaker. Then, the amount of water that is in the beaker is measured. The investigation will determine which type of soil retains the most water.

Water Filtered Through Types of Soil

Type of Soil	Amount of Water Filtered
Clay	41 mL
Loamy	72 mL
Sandy	97 mL

1. Fill in each blank with the word that correctly completes the sentence.

hypothesis	variable	weakness

The type of soil is a ____________________ in the study.

Completing each trial only one time is a ____________________ of the study.

A ____________________ could be that loamy soil retains more water than sandy or clay soil.

2. Which step would come first in completing this scientific investigation?

   A. identifying the question/topic being investigated

   B. creating a hypothesis as to which type of soil will retain the most water

   C. forming a conclusion based on evidence and results shown in the data table

   D. measuring and recording the data of how much water is filtered through the soil

3. Which type of soil retained the least amount of water?

   A. clay

   B. loamy

   C. sandy

4. What conclusion can be drawn from the data in the table?

   A. Sandy soil is best for growing plants.

   B. Loamy is a mixture of sandy soil and clay.

   C. Plants grown in clay will be taller than other plants.

   D. The amount of water retained by the clay is the greatest.

5. Which symbol correctly completes the sentence?

   The amount of water retained by the sandy soil is ________ the amount of water retained by the loamy.

   A. $>$

   B. $<$

   C. $\geq$

   D. $=$

6. Dionte makes a mixture of clay and loamy. Which is the most accurate prediction of how much water will be filtered through the mixture?

   A. 35 mL

   B. 42 mL

   C. 55 mL

   D. 80 mL

Use the following information to answer questions 7 and 8.

A classroom is surveyed to find the number of solids, liquids, and gases within the room. The results are shown in the following table.

Matter Observed in Classroom	
Property of Matter	**Frequency Observed**
Gas	1
Liquid	5
Solid	24

7. Which statement accurately describes the results of this investigation?

   A. Fewer solids than liquids were present.

   B. Very few gases are present in the world.

   C. In this room, there were more solids than either liquids or gases.

   D. The number of solids in a room will always outnumber the liquids in a room.

8. What is a weakness of this investigation?

   A. Fewer than 25 items were included.

   B. The results of the study are limited to the room studied.

   C. The terms *solid*, *liquid*, and *gas* were not defined for the reader.

   D. The frequency of solids, liquids, and gases were unequal.

Fill in the blanks.

9. The term ________________________ describes all the chemical reactions that must occur for an organism to survive.

10. If you consume more calories than you use, they will be ________________________ as fat or glycogen.

11. Yeasts convert sugar into carbon dioxide during the production of bread; they do this through the process of ________________________.

12. When you get a bacterial infection, your immune system sees the antigens on the bacteria as foreign and produces ________________________ against them.

13. During a solar eclipse, ________________________ moves between Earth and the sun.

14. ________________________ weathering breaks rocks into smaller pieces without changing their chemical composition.

Select the correct option.

15. Which of the following is NOT a simple machine?

    A. inclined plane

    B. wedge

    C. lever

    D. chain

16. The code of ________ is transcribed to ________ so that ________ can be made.

    A. RNA, DNA, proteins

    B. DNA, RNA, proteins

    C. RNA, DNA, amino acids

    D. DNA, RNA, amino acids

17. ________ are substances that we take in from the food we eat. They help reactions occur in our bodies.

    A. Fats

    B. Calories

    C. Vitamins

    D. Carbohydrates

18. The equation for the burning of methane is balanced. This means that in the products is the same number of each type of (atom, molecule, compound) as in the reactants.

19. During a reaction, more heat is released by the formation of bonds in the products than is absorbed by breaking the bonds in the reactants. This tells us the reaction is (endothermic, exothermic).

Use the following information to answer questions 20 and 21.

In tomatoes, round fruit (R) is dominant to long fruit (r). In an experiment, a dominant plant (RR) is crossed with a recessive plant (rr).

20. Which best describes the possible offspring?

    A. 25% grow round fruit, 75% grow long fruit

    B. 50% grow long fruit, 50% grow round fruit

    C. 25% grow long fruit, 75% grow round fruit

    D. 100% grow round fruit

21. What will the phenotypes of the offspring be?

    A. 50% Rr, 50% RR

    B. 25% long, 75% round

    C. 25% RR, 50% Rr, 25% rr

    D. 100% round

Answer the following questions.

22. Explain why human sperm and egg cells have 23 chromosomes.

____

____

23. What would happen to the temperature of Earth's surface if global ocean currents stopped?

____

____

24. Which type of electromagnetic radiation has more energy than visible light waves?

    A. ultraviolet waves **C.** microwaves

    B. radio waves **D.** infrared waves

25. Which of these is an agent of chemical weathering?

    A. wind **C.** ice

    B. water **D.** plant roots

26. The air coming in through a vent is about 78% nitrogen gas (N_2) and 21% oxygen gas (O_2), as well as some carbon dioxide (CO_2) and water gas (H_2O). Which best describes what the air is?

    A. element **C.** homogeneous mixture

    B. compound **D.** heterogeneous mixture

27. Explain how you know the burning of methane is a chemical change.

____

____

____

28. Which of these is NOT a landform?

    A. a river valley

    B. an asteroid impact crater

    C. a beach

    D. a city

29. Which layers of Earth are made of solid rock?

    A. crust, mantle, and outer core

    B. crust, top portion of mantle, and inner core

    C. top portion of mantle, outer core, and inner core

    D. bottom portion of mantle, outer core, and inner core

30. Which of these objects in our solar system is made of gas?

    A. the moon **C.** Jupiter

    B. Mars **D.** Venus

31. Trade winds that blow counterclockwise are found in what part of the globe?

    A. northern hemisphere **C.** near the equator

    B. southern hemisphere **D.** near the poles

32. A homeowner decides to fertilize trees with nitrogen-based fertilizer. He adds more fertilizer than the trees can use. That year, the trees grow faster. What are the probable effects on the carbon and nitrogen cycles?

    A. The trees will add more carbon into the atmosphere because they are growing faster, and excess nitrogen in the soil will be taken up by the atmosphere.

    B. The trees will add more carbon to the soil because they are growing faster, and excess nitrogen will be taken up by nitrogen-fixing bacteria in the soil.

    C. The trees will take up more carbon from the atmosphere because they are growing faster, and excess nitrogen will run off into nearby water, disrupting the ecosystem.

    D. The trees will take up more carbon from the atmosphere because they are growing faster, and excess nitrogen will be taken up by nitrogen-fixing bacteria in the soil.

33. How does plant respiration contribute to the carbon cycle?

    A. Carbon moves from the plant into the soil.

    B. Carbon moves from the plant into the atmosphere.

    C. Carbon moves from the atmosphere into the plant.

    D. Carbon moves from the plant into an animal.

34. A scientist measures the amount of a particular radioactive material in a rock and compares it to the decay rate of that material. What does this tell the scientist about the rock?

    A. how old it is

    B. what it is made from

    C. what planet it came from

    D. how it was formed

35. Which of the following is evidence that the universe is expanding?

    A. Scientists observe stars being born and other stars dying.

    B. Scientists count hundreds of billions of galaxies in the universe.

    C. Scientists observe that stars and galaxies are all moving away from each other.

    D. Scientists observe that light from distant galaxies is getting brighter.

Read the following passage and study the graph shown. Then, answer questions 36–39.

Reindeer can be found across the Arctic and in Europe, Asia, and North America. In 1911, 25 reindeer were brought to one of the Pribilof Islands off the coast of Alaska, where they have no natural predators. The graph below shows the population of these reindeer from 1911 to 1950.

Reindeer Populations, 1911–1950

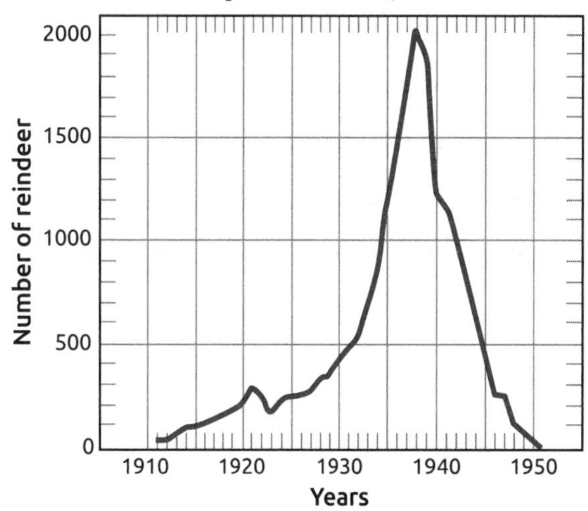

© American Association for the Advancement of Science, 1951

36. The overall population of reindeer ______________________ between the years 1911 and 1950.

    A. increased

    B. decreased

    C. increased then decreased

    D. decreased then increased

37. How might the presence of a predator have affected the population changes observed?

____

____

____

38. Which of the following summarizes what most likely happened to the reindeer on the Pribilof Islands?

   A. The reindeer killed one another fighting for mates.

   B. The reindeer exceeded the carrying capacity of the island.

   C. The reindeer left the island in search of better feeding grounds.

   D. The reindeer could not adapt to the cold conditions and died off.

39. Wolves chase after reindeer, only catching those that are slower than the rest. What effect might this have on the wolf population?

   A. The wolves will become extinct.

   B. The wolves will become stronger.

   C. The wolves will become plant eaters.

   D. The wolves will become prey for the reindeer.

Use the following information to answer questions 40–43.

A caterer is using a pulley to lift a block of ice off the ground so he can put a cart under it to move it to his freezer. The block of ice has a mass of 75 kg, and the maximum height it can be raised with the pulley (point Y) is 3 meters above the ground (point X).

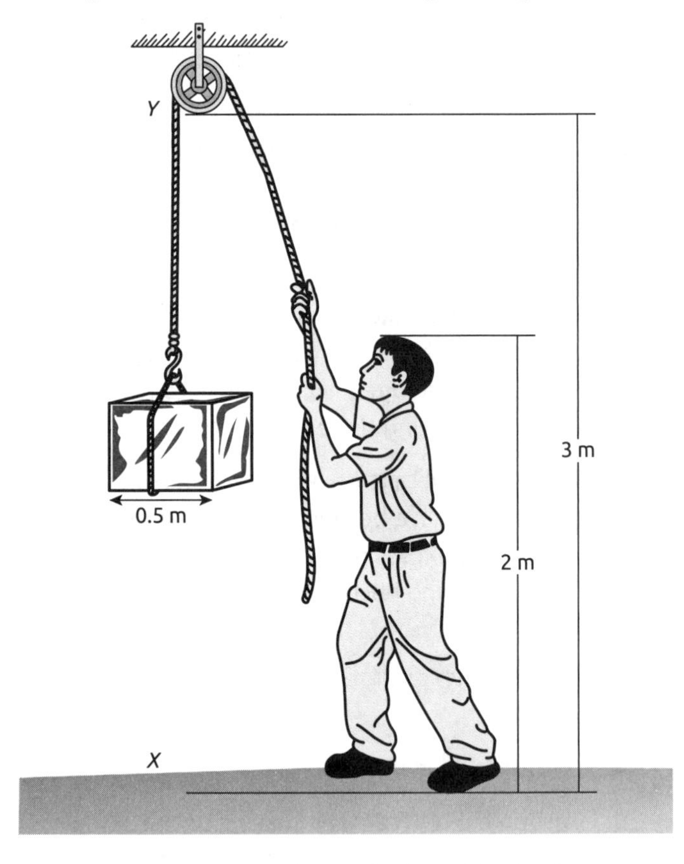

40. When the caterer has pulled the block all the way to point Y, the block can't move any farther. At point Y, the potential energy of the block is (more than, less than, about the same as) the kinetic energy of the block, and when the block is on the ground at point X, the potential energy of the block is (more than, less than, about the same as) the kinetic energy of the block.

41. The caterer pulled the ice block from point X to point Y in 6 seconds, at an average velocity of ____________________ m/s upwards.

42. When the ice block was at point Y, the rope slipped through the caterer's hands, and the block fell. The acceleration of the block toward the ground was 9.8 m/s^2. What was the velocity of the block after 0.2 seconds?

A. 0.4 m/s

B. 2.0 m/s

C. 4.9 m/s

D. 9.8 m/s

43. The mass of the cart is 25 kg. In order to give the cart with the ice block an acceleration of 5 m/s2, the caterer had to apply a force of

_______________________ N.

Use the following information to answer questions 44–45.

Water can be heated for household use by burning natural gas, as shown in the diagram. As hot water is used, cold water refills the tank. The tank is heated from below by the burning natural gas. The vent supplies air from the outside and is also how the gaseous products from the burning gas escape. The reaction for natural gas burning is:

$CH_4(g) + 2O_2(g) \rightarrow CO_2(g) + 2H_2O(g)$

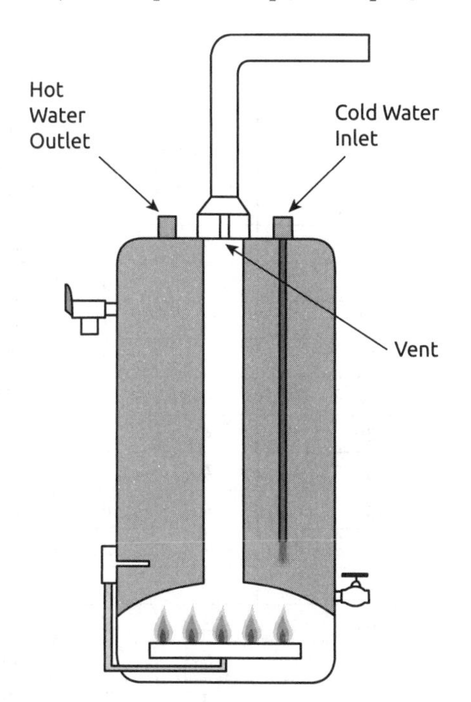

44. Compare using natural gas to heat water with using solar power to heat water.

____

____

Draw a line from each phrase to the word it describes.

45. **A.** how heat travels from the flames to the metal wall of the tank — conduction

B. how heat travels from the metal wall of the tank to the water — convection

C. how heat travels from the water at the bottom to the water at the top of the tank — radiation

ANSWER KEY

1. variables, weakness, hypothesis
2. A.
3. C.
4. D.
5. B.
6. C.
7. C.
8. B.
9. metabolism
10. stored
11. fermentation
12. antibodies
13. the moon
14. Physical
15. D.
16. B.
17. C.
18. atom
19. exothermic
20. D.
21. D.
22. Sperm and egg cells fuse to create offspring that have 46 chromosomes.
23. Sample answer: Areas near the equator would be much warmer, and areas near the poles would be much colder.
24. A.
25. B.
26. C.
27. Sample answer: When a new substance is formed, a chemical change has occurred.
28. D.
29. B.
30. C.
31. B.
32. C.
33. C.
34. A.
35. C.
36. C.
37. Sample answer: A predator would have made the increase in reindeer smaller, but the population most likely would have leveled off over time.
38. B.
39. B.
40. more than; less than
41. 0.5
42. B.
43. 500
44. Sample answer: Natural gas is less expensive than solar power, but solar power is renewable, while natural gas is not. Natural gas produces carbon dioxide, which is a greenhouse gas, while solar power does not.
45. A. radiation; B. conduction; C. convection

Check your answers. Review the questions you did not answer correctly. You can use the chart below to locate lessons in this book that will help you learn more science content and skills. Which lessons do you need to study? Work through the book, paying close attention to the lessons in which you missed the most questions. At the end of the book, you will have a chance to take another test to see how much your score improves.

Question	Where to Look for Help		
	Unit	Lesson	Page
1, 2	1	1	13
3, 4, 6, 7, 8, 36	1	4	19
5	1	3	16
9	2	1	25
10, 11	2	9	54
12	2	8	52
13, 30	4	7	135
14, 27	3	9	100
15	3	7	93
16, 22	2	3	32
17	2	10	56
18	3	10	103
19, 45	3	3	80
20, 21	2	4	36
23, 31	4	2	120
24	3	4	83
25	4	4	126
26	3	8	96
28, 34	4	5	129
29	4	3	123
32, 33	4	1	116
35	4	6	132
37, 38, 39	1	3	18
40	3	1	72
41, 42	3	5	86
43	3	6	89
44	3	2	76

Science Reasoning

Some people say that children ask more than 300 questions a day. Adults ask questions as well, but probably not as many. Asking questions helps us learn and find answers. Some questions are answered easily, while others require research. Science can help us find the answers to many of the questions we ask. Questions lead to predictions, and predictions can be tested by experiments. Through these investigations, scientists are able to answer questions and teach the world what they learn.

Unit 1 Lesson 1 INVESTIGATION DESIGN

Scientific investigations have led to the discovery of electricity. They have led to cures for diseases. American physicist Brian Greene said, "Science is the process that takes us from confusion to understanding." Investigations are based on observations. They follow specific steps to test a question. Observational investigations use the senses to gather data.

Key Terms

control variable

dependent variable

hypothesis

independent variable

Scientific Investigation

All investigations begin with asking a question and developing a hypothesis. A **hypothesis** is a prediction or an educated guess that answers a question. For example, say we are trying to find which toy car will travel the farthest distance down a ramp, car A, B, or C. The question is, "Which car will travel the farthest distance when released down a ramp?" The hypothesis might be that car B will travel the farthest distance.

Experiments have variables. Variables are factors that can be changed to different amounts during experiments. **Independent variables** are changed by the scientist during an experiment. **Dependent variables** change in response to the changing independent variable. **Control variables** are kept constant, or the same, on purpose. In our investigation, the car is the independent variable. The distance traveled is the dependent variable. Control variables include factors such as the height, length, and angle of the ramp.

The illustration below shows the setup of the car investigation.

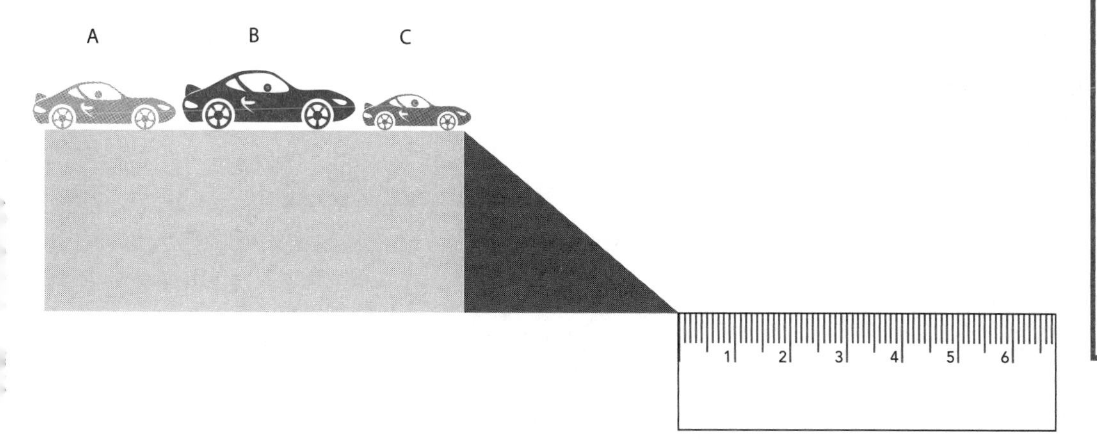

Real-World Connection

A scientist grows bean plants in sandy, clay, or loamy soil to learn which soil helps the plants grow best. The independent variable is the type of soil. The dependent variable is the height of the plant. A control variable is the type of plant.

LESSON REVIEW

Complete the activities below to check your understanding of the lesson content. The Unit 1 Answer Key is on page 152.

Skills Practice

Use the table to answer the questions.

An investigation was done to test how metals react with water at a temperature of 25°C. The results are shown in the data table.

Results of Metal and Water	
Sample	**Reaction with water after 1 minute**
Metal A	No change
Metal B	Colorless gas formed quickly.
Metal C	Bubbles formed slowly on surface.

1. Which is a possible hypothesis for the above investigation?

   A. How do metals react with water?

   B. Which metals are used in the investigation?

   C. All the metals will produce bubbles when they react with water.

   D. The results show that Metal A had no reaction with the water.

2. Which is the independent variable in the above investigation?

   A. the types of metal used

   B. the temperature of the water

   C. the changes observed in metal B

   D. the reactions of the metals and water

3. Which is a control variable in the above investigation?

   A. the types of metals used

   B. the temperature of the water

   C. the reaction between metal C and water

   D. the type of gas formed from metal B and water

EVALUATION OF INVESTIGATIONS

The scientific method is a series of standard steps. It is important in a scientific investigation to stick to the facts and not let your opinions get in the way. This helps keep the investigation free of bias and minimize error.

Key Terms

bias

double blind

error

Evaluating an Investigation

Bias is anything that sways an experiment's results in a way that makes them inaccurate. A bias can come from a wish for personal gain. It can also result from a bad design for the experiment or a belief that the hypothesis is true before the experiment is done.

Errors are mistakes. Measuring instruments are never 100 percent accurate, and sampling techniques are not perfect. So, error is expected. Error percentages are calculated and included in data. In a scientific investigation, it is important to be exact and thorough. Scientists try to keep error rates as low as possible.

In some medical research, the investigations are designed to be **double blind**. Some patients in the investigation will be given medication and others are given a placebo. A placebo is a pill that looks like the medication but contains no medicine. Neither patients nor investigators know which patients receive which pill. After the results of the study are completed, this information will be revealed. This keeps the results free of bias. The results found accurately depict the difference the medication makes.

Strengths and Weaknesses

An investigation must be reviewed carefully. It should be detailed enough that it can be repeated by others. It should be evaluated for its strengths and weaknesses. A strength might be the large number of trials that were done to get accurate results. Another strength could be the careful calibration of measuring instruments. A weakness might be poorly controlled variables, such as wind or temperature, that could skew the results.

The table below shows the results of the car and ramp investigation from Lesson 1.

Distance Cars Traveled (in inches)			
	Trial 1	**Trial 2**	**Trial 3**
Car A	5.75 in	6.0 in	4.5 in
Car B	7.25 in	8.5 in	14.0 in
Car C	6.5 in	5.25 in	6.75 in

What could be some errors or weaknesses found in this investigation? The distance traveled in trial 3 for car B is very different from the other trials. The investigator should consider why that might be. Maybe he accidentally gave the car a slight push as it was released. A strength is that repeated trials were completed. It is important to repeat investigations to be sure results are correct.

Real-World Connection

In 1911, fossils of a skull were found that were thought to link apes and humans. The organism the skull supposedly belonged to was named Piltdown Man. In 1953, Piltdown Man was exposed as a hoax. It turned out the fossils had been altered to prove the scientists' hypothesis. This is a well-known and extreme example of bias in science.

LESSON REVIEW

Complete the activities below to check your understanding of the lesson content. The Unit 1 Answer Key is on page 152.

Vocabulary

Write definitions in your own words for each of the key terms.

1. bias ____________________

   ____

2. double blind ____________________

   ____

3. error ____________________

   ____

Skills Practice

Answer the questions based on the content covered in the lesson.

4. An investigation was done testing how metals react with water at 25°C. The results are shown in the data table.

Results of Metal and Water	
Sample	**Reaction with water after 1 minute**
Metal A	No change
Metal B	Colorless gas formed quickly.
Metal C	Bubbles formed slowly on surface.

In the investigation described above, which information does NOT need to be included to be sure it can be repeated?

A. the amount of water

B. the amount of metals

C. the temperature of the water

D. descriptions of the reaction of the metals with the water

5. Several students threw paper airplanes at the same time and measured the distances flown. This was repeated 5 times. Which could be a possible source of error in this investigation?

A. The students threw the planes 5 times.

B. The students used the same type of paper to build the planes.

C. Different amounts of force can be used to throw the planes.

D. The wind force and direction might have affected the distance the planes flew.

COMPREHENDING SCIENTIFIC PRESENTATIONS

Scientists often represent data in visual ways. Using words alone can sometimes be confusing. Scientists often put their data in graphs, charts, tables, and diagrams. This helps them better understand and interpret their observations.

Key Terms

- chart
- data
- diagram
- graph
- table

Making Comparisons

Take, for example, the comparison of precipitation in two areas of the country: Seattle, Washington, and the Twin Cities in Minnesota.

The following image uses both a **graph** and a **table** to compare the **data**. The key shows you how to differentiate the data in the graph. Here, light purple represents Seattle, and dark purple represents the Twin Cities.

Annual Precipitation Rates for Twin Cities and Seattle

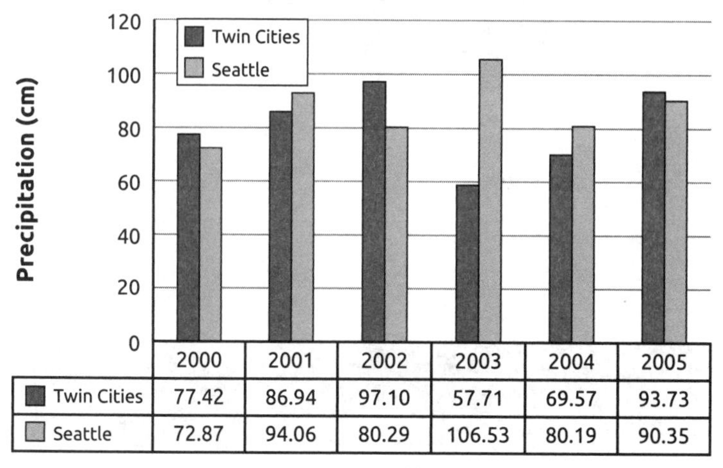

	2000	2001	2002	2003	2004	2005
Twin Cities	77.42	86.94	97.10	57.71	69.57	93.73
Seattle	72.87	94.06	80.29	106.53	80.19	90.35

Year

Source: U.S. Department of Transportation

In the top portion, the precipitation of the two areas is shown as a graph. You can see which city had more or less rainfall in a particular year. For example, you can see that in 2003 the lighter colored bar is much higher than the colored bar. This means that Seattle had more precipitation that year.

The bottom rows in the table show the amount of precipitation. You can look at the table and see the exact numbers recorded. For example, in 2002 the Twin Cities had 97.10 cm of precipitation. Seattle had 80.29 cm of precipitation.

Charts, Tables, and Diagrams

Charts and tables show data as numbers or in words. A **diagram** is usually a picture used to represent something real. For example, a diagram of the human heart can be an illustration showing the different parts.

COMPREHENDING SCIENTIFIC PRESENTATIONS

Graphs

Graphs have two axes, the *y*-axis and the *x*-axis. On the graph on page 16, "Precipitation" is on the *y*-axis and "Year" is on the *x*-axis. The axes should always be labeled, and the graph should always have a title.

There are several different types of graphs, such as the bar graph used on the previous page. Scientists also utilize other types of graphs, including line graphs and pie charts.

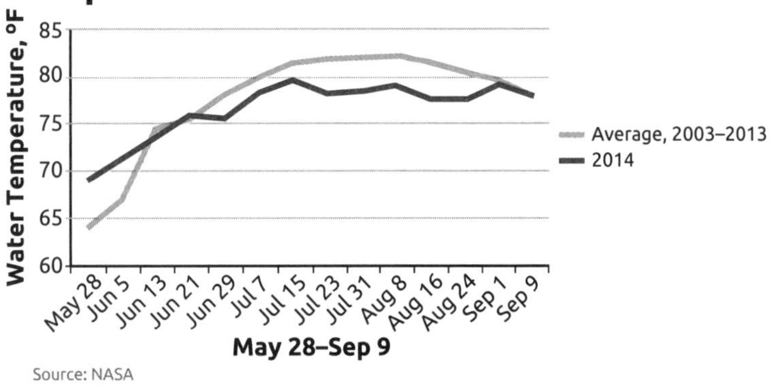

Line graph

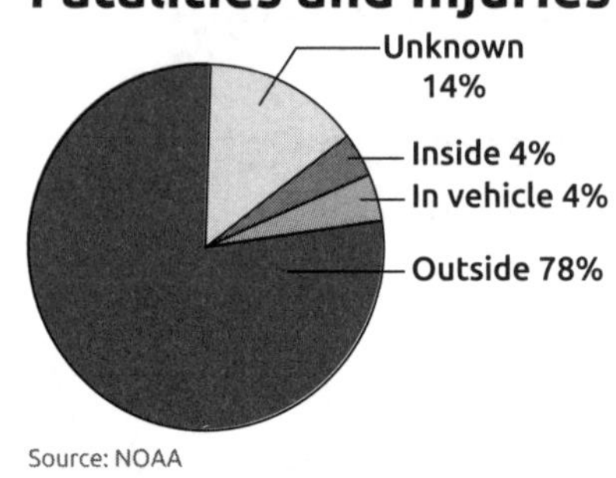

Pie chart

LESSON REVIEW

Complete the activity below to check your understanding of the lesson content. The Unit 1 Answer Key is on page 152.

Skills Practice

Answer the questions based on the content covered in the lesson.

1. According to the line graph in the lesson, in which week in 2014 did Chesapeake Bay have the highest surface water temperature? ____________________

2. According to the pie chart in the lesson, where did the most lightning injuries and fatalities in New Mexico take place? ____________________

REASONING FROM DATA

Dr. Lewin has just completed an investigation in a forest ecosystem that is being cut down. She has observed how the number of prey in the ecosystem affects the number of predators. She carefully considers her data and then comes to the conclusion that the fewer prey there are in the food web, the fewer predators there are. She then predicts that if the number of prey continues to decrease, the number of predators will decrease as well. How did Dr. Lewin use her data to form a conclusion and a prediction?

Key Terms

conclusion

evidence

model

prediction

Making Conclusions

Dr. Lewin put her data into a table:

Month	Prey	Predators
1	2,010	280
2	1,914	269
3	1,815	258
4	1,709	246
5	1,611	234
6	1,515	226

By looking at the data in her table, she was able to see that as the number of prey was decreasing, so was the number of predators. Therefore, she was able to form a **conclusion**. She concluded that as the number of prey decreases, the number of predators decreases. She used **evidence** from her investigation to support this conclusion. For example, the prey decreased from 1,914 to 1,815 in month 3. The predators decreased from 269 to 258 in the same month.

Forming Predictions

After using her data to come to a conclusion, Dr. Lewin was able to make a **prediction**. She used the data to make a prediction about the ecosystem. She concluded that for every 100 fewer prey, there will be about 11 fewer predators. She could use this information to make a prediction about month 7. She could predict that if the prey decreases to 1,414 in month 7, the predators will decrease to about 215.

Scientists use different types of data to make predictions about future events. They can use direct observations. They can also use **models**. A model makes something that is difficult to observe easier to understand. For example, atoms are too small to be seen with the eye alone. Scientists can use a model to show how atoms will react with each other.

Real-World Connection

Scientists often use computer models to predict the weather. The computer models show different scenarios of what could happen. The scientists make forecasts based on these predictions.

LESSON REVIEW

Complete the activities below to check your understanding of the lesson content. The Unit 1 Answer Key is on page 152.

Skills Practice

Answer the questions based on the content covered in the lesson.

1. How do scientists make conclusions?

____

____

2. How are models useful to scientists?

   A. They can be used to organize data.

   B. They can show actual events in the future.

   C. They can make information easier to understand.

   D. They can determine the best steps to take in an investigation.

3. Nate heats a gas inside a closed container. He then takes measurements of both the temperature and pressure inside the container. He plots his data on a graph, as shown below.

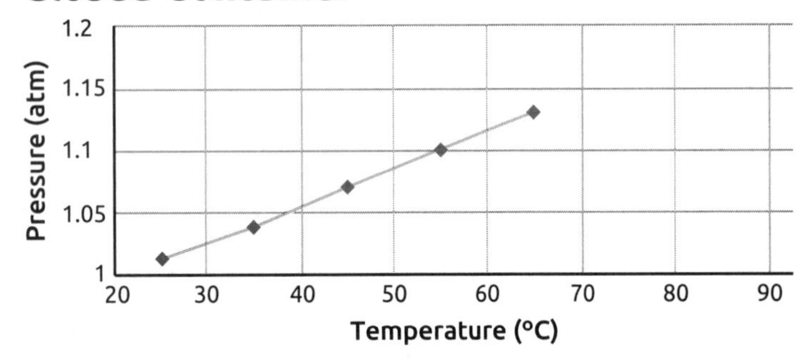

What can Nate conclude about his investigation? What prediction can he make about the pressure of the gas at 85°C?

____

____

EXPRESSING SCIENTIFIC INFORMATION

A scientist measured the energy from the sun in China over a period of one year. He recorded his data in the graph below.

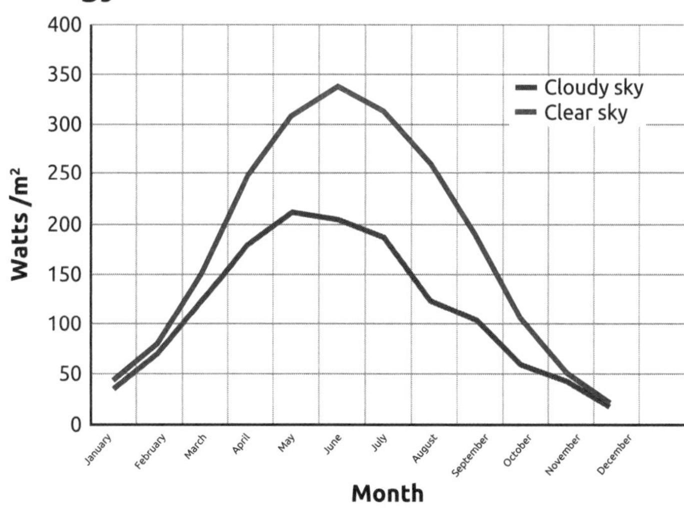

Source: NASA

Graphs and tables are useful for showing data in scientific investigations. However, it is important to be able to describe the data in other ways. How could the scientist describe his data about the sun's energy differently?

Using Numbers

Data can be described using numbers. Look at the graph above. In June and July, the energy China received from the sun was around 340 watts/m^2 on clear days; on cloudy days, it was about 210 watts/m^2. Sometimes, as in the example just used, you need to estimate numbers. The interval for the *y*-axis is 50 watts/m^2, but the data falls between these intervals.

Using Words

Data can also be described using words. Using the interval of June and July again, you can say that China receives more energy from the sun on clear days than on cloudy days. You can also say that the energy China receives from the sun is highest on clear days.

Using Symbols

Sometimes data is described using symbols. The most commonly used symbols are as follows:

> greater than	< less than	= equal to
≥ greater than or equal to	≤ less than or equal to	≠ not equal to

Complete the activities below to check your understanding of the lesson content. The Unit 1 Answer Key is on page 150.

Skills Practice

Use the following graph to answer the questions.

A group of engineers tested how high the X-99 could fly during a flight test. After four trials, they plotted their data in a bar graph.

X-99 Flight Test Results

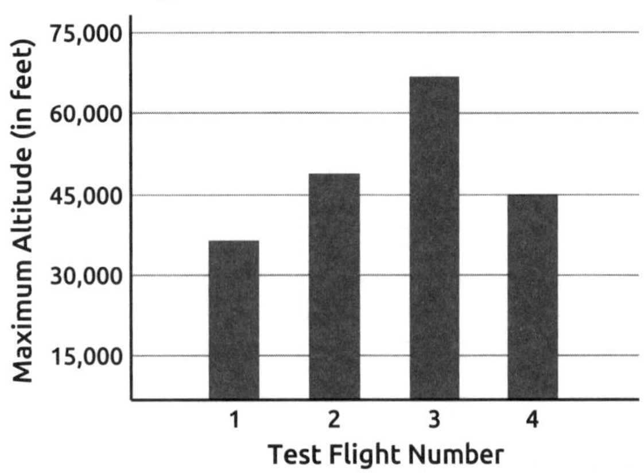

Source: NASA

1. What was the maximum altitude in test flight 4? ____________________

2. Which test flight reached the highest altitude? ____________________

3. Which symbol completes the sentence?

   The maximum altitude of test flight 2 was ________ the maximum altitude of test flight 1.

   A. <

   B. >

   C. ≤

   D. ≥

UNIT PRACTICE TEST

Answer the questions based on the content covered in this unit. The Unit 1 Answer Key is on page 152.

Use the following information about an investigation to answer questions 1–3.

A scientist wants to determine if pill bugs prefer dark or light locations. She puts a desk lamp over one end of a table. The opposite end of the table is shaded and dark. The scientist releases 25 pill bugs in the middle of the table and then records their locations after 2, 4, 6, and 8 minutes.

1. Fill in the blank with the word that correctly completes the sentence.

   The location of the pill bugs is the

   ______________________ variable.

2. Draw a line to match the term to the example from the pill bug investigation.

Term	**Examples**
question/topic	75% of the pill bugs will stay in the dark.
hypothesis	the time increments measured
independent variable	the amount of light
control variable	Do pill bugs prefer dark or light locations?

3. What could be an error in the pill bug investigation?

   ____

   ____

Use the following information about an investigation to answer question 4.

A study is being conducted about tongue rolling. The researcher believes that 75 percent of children can roll their tongues. He asks a group of 5 children to show him if they can roll their tongues. The results are recorded.

4. What is a weakness of this investigation?

   ____

Use the following graph to answer questions 5 and 6.

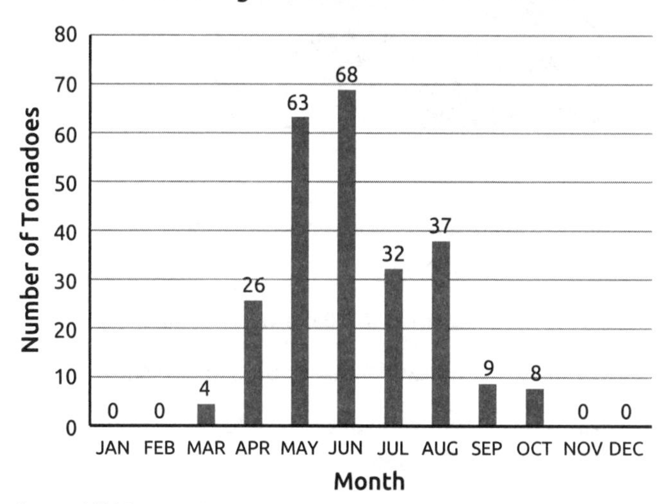

Source: NOAA

5. Which month had the greatest number of tornadoes?

   A. April

   B. May

   C. June

   D. July

6. What conclusion can be drawn from the data in the graph?

   A. The same number of tornadoes occurred each year from 1995 to 2013.

   B. Tornadoes are most likely to occur in the summer months.

   C. The frequency of tornadoes increases each year.

   D. There are no tornadoes in the fall and spring.

Unit 1 Lessons 1–5

UNIT PRACTICE TEST

Use the following information about an investigation to answer questions 7–10.

Terrence conducts an investigation to test the solubility of sugar and salt. He observes how much of each substance dissolves in the same amount of water at different temperatures. He then records his observations in the chart below.

Temperature (°C)	0	20	40	60	80	100
Salt (g)	33.0	33.5	34.0	35.0	35.5	36.5
Sugar (g)	172.0	197.0	236.0	285.0	359.0	486.0

7. Which statement most accurately describes the result of this experiment?

   A. More sugar dissolves in water than salt.

   B. Both sugar and salt dissolve at the same rate.

   C. As the temperature increases, the amount of salt dissolved decreases.

   D. The amount of sugar dissolved in water is equal to the amount of salt dissolved in water.

8. What can Terrence conclude about the solubility of sugar as compared to the solubility of salt?

   ____

   ____

9. Terrence wants to predict how much salt would dissolve at 120°C. Which is the most accurate prediction?

   A. 32.5 g

   B. 35.0 g

   C. 36.5 g

   D. 37.0 g

10. Which symbol correctly completes the sentence? At 20°C, the amount of salt dissolved is ________ the amount of sugar dissolved.

    A. $>$

    B. $<$

    C. $\geq$

    D. $=$

Life Science

You wake up in the morning and prepare a bowl of cereal. As you're enjoying your breakfast, you begin to wonder how the milk and cereal provide you with energy during the morning hours. You breathe in the fresh air and think about all the changes that happen in your body to keep you healthy and productive throughout the day.

Biologists are fascinated with similar questions. They seek to find out how living beings function. Using microscopes in labs, they explore how our bodies break down the food we eat, repair themselves when we get injured, and fight off diseases.

There are basic differences between living organisms and inorganic matter. All living organisms are born. Eventually, they all die. They are capable of creating another one of their own species. And, they have five basic needs: water, air, light, food, and the correct temperature. Some living organisms can survive only in a hot environment and others can tolerate only cold temperatures.

Unit 2 Lesson 1

ESSENTIAL FUNCTIONS AND COMPONENTS OF LIFE

Cells

The **cell** is the basic unit of life. Every living organism is made of one or more cells. A cell contains multiple functional parts that are held together with a cell membrane. To support the various functions within a cell, the cell membrane allows nutrients to enter the cell and passes the waste outward. Plant cells have an extra layer outside the membrane, called a cell wall. All cells share basic properties and are also individually unique. For example, all cells have a nucleus that contains the genetic information—or DNA—of the cell.

Key Terms

cell

chemical reaction

enzyme

metabolism

Chemical Reactions

Thousands of chemical reactions take place in the cells of all living organisms. In a **chemical reaction**, the atoms in one or more substances rearrange and change how they are connected, producing at least one new substance. The substances that change are called reactants. The new substances created are called products. The reaction also either consumes or releases energy, depending on the reactants and products. For example, glucose reacts to produce carbon dioxide and water, releasing energy. Our cells use this reaction to gain energy to use in other reactions. Different cells use different reactants and create different products.

ESSENTIAL FUNCTIONS AND COMPONENTS OF LIFE

Metabolism

Metabolism is the sum of all chemical reactions that take place within a living organism to maintain life. During metabolism, food particles are broken down by the cell to make energy and build new materials for cell needs. As a result of metabolism, waste products are formed in the cell. The cell membrane passes these waste products out of the cell, where they can then be removed from the body of the living organism.

There are two types of metabolism: catabolism and anabolism. In catabolism, the cell breaks down larger particles into smaller ones. Some energy is released as a result of this process. In anabolism, the cell combines smaller particles to make larger ones. In this process, energy is consumed.

Enzymes

An **enzyme** is a large molecule that speeds up a chemical reaction within a living organism. Enzymes play an essential role in the function of all organs, tissues, and cells. A target molecule, called a substrate, attaches itself to the enzyme. Once the reaction takes place, the products are detached from the enzyme, and another substrate is ready to attach itself again. For example, digestive enzymes help break down food particles much more quickly than they would break down on their own.

Real-World Connection

The shape of an enzyme changes with temperature. If the enzyme is too hot or too cold, it will not be the right shape to correctly attach to a substrate. One of the dangers of a fever is that the high temperature may prevent enzymes in the body from performing their job.

Enzyme Breakdown

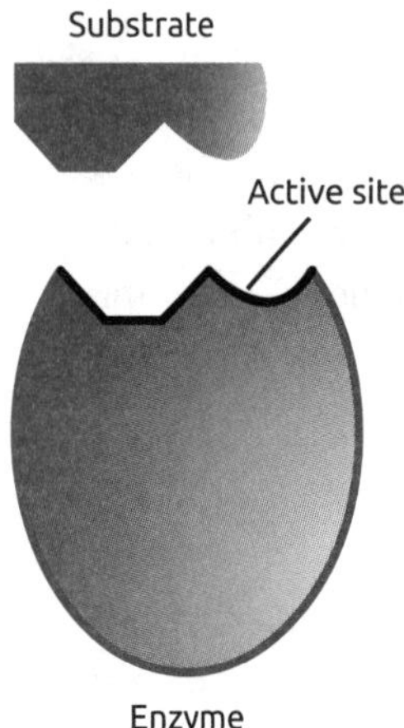

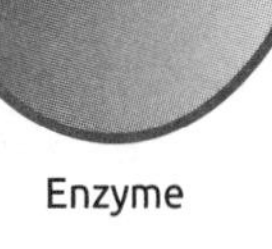

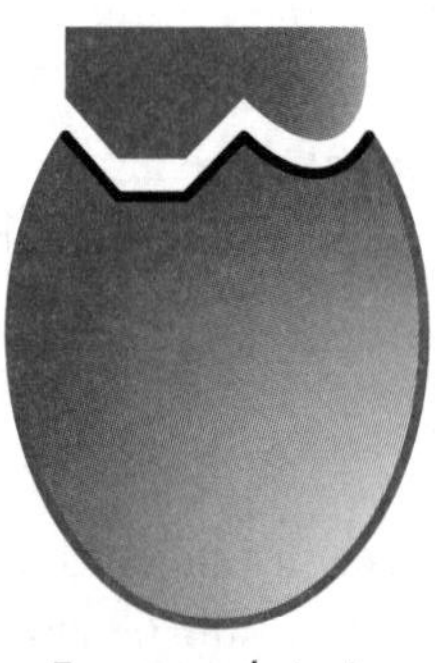

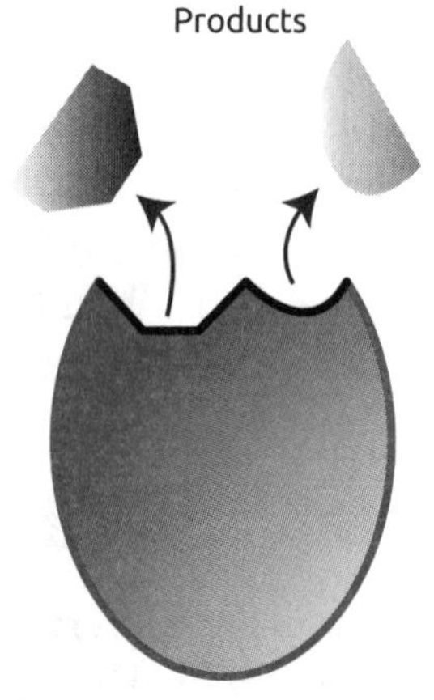

LESSON REVIEW

Complete the activities below to check your understanding of the lesson content. The Unit 2 Answer Key is on page 152.

Vocabulary

Write definitions in your own words for each of the key terms.

1. cell ______________________________

____

2. chemical reaction ______________________________

____

3. enzyme ______________________________

____

4. metabolism ______________________________

____

Skills Practice

Answer the questions based on the content covered in the lesson.

5. Select ALL the options that are needed for all living organisms.

   A. water

   B. fire

   C. food

   D. right temperature

6. Which of the following is an example of a chemical reaction?

   A. Sugar dissolves in water and produces sweet water.

   B. Sunlight melts the snowflakes on the ground.

   C. Carbon combines with oxygen and produces carbon dioxide.

   D. Water boils and evaporates into steam.

Fill in the blanks.

7. All cells have a ____________________, which contains the genetic information—or DNA—of the cell.

8. The new substance created as a result of a chemical reaction is called a ____________________.

9. A cell contains multiple functional parts that are held together with a ____________________.

10. During ____________________, the cell breaks down larger particles into smaller ones. As a result of this type of cellular metabolism, some energy is released.

CELL THEORY AND GROWTH

Key Terms

cell cycle

chromosome

meiosis

mitosis

organ

organelle

system

tissue

Many functions take place in a cell to maintain the health of the cell and the living organism. Each human cell is specialized to perform a certain task. They work together to perform every function in our bodies.

Organelle

All cells contain smaller structures, called **organelles**, that have special functions. Organelles are supported by a jelly-like substance in the cell called the cytoplasm. An important set of organelles responsible for producing energy are called mitochondria. Mitochondria are the powerhouses of the cell and produce the energy it needs. The nucleus is another example of an organelle. The nucleus has two major roles: it protects the cell's DNA, and it controls the activities of the cell, such as growth, metabolism, and reproduction.

The DNA molecule, which contains the instructions for all the cell's activities, is packed inside a thread-like structure called a **chromosome**. Chromosomes contain special proteins that help keep the DNA coiled and well wrapped, making them an efficient way of storing the DNA. Chromosomes are not visible, even under a microscope, unless the cell is dividing. Each human cell has 46 chromosomes placed in a set of 23 pairs.

Animal Cell

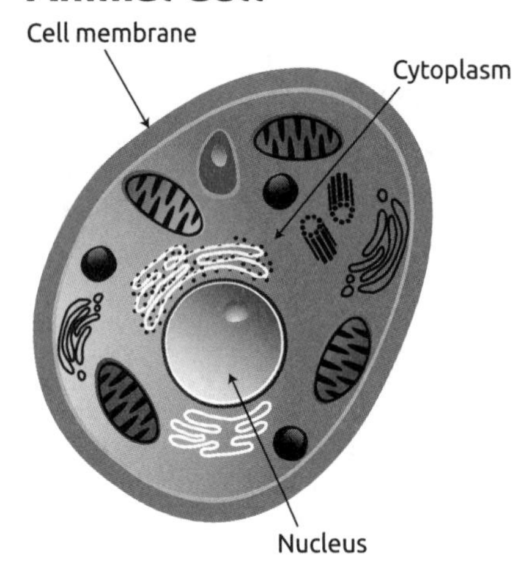

Mitosis

Mitosis is a form of cell division by which a single cell divides into two identical cells. The two resulting cells are called daughter cells. The new cells may also divide to make new daughter cells. The body uses mitosis to produce new cells for growth or repair of damaged or old cells. The process of division of a cell into new identical cells from the beginning to the end is called the **cell cycle**.

During mitosis, the nucleus produces identical pairs of chromosomes. Then, the cytoplasm of the cell begins to divide so that each daughter cell has a nucleus surrounded by cytoplasm. Most of the cells in the body divide through mitosis. The entire process of mitosis takes place in four stages: prophase, metaphase, anaphase, and telophase.

Meiosis

Meiosis is another form of cell division by which a single cell divides into four daughter cells. Unlike mitosis, each of the daughter cells resulting from meiosis has only half of the chromosomes of the original cell—one from each chromosome pair. Meiosis is an eight-step process, and it takes places only during the production of egg cells and sperm cells.

Eggs and sperm have half the normal number of chromosomes because they are combined during reproduction. When they are combined, the new cell will have a full set of chromosome pairs. The new cell can now multiply by mitosis and grow. In the case of humans, the fertilized cell will eventually become a human baby.

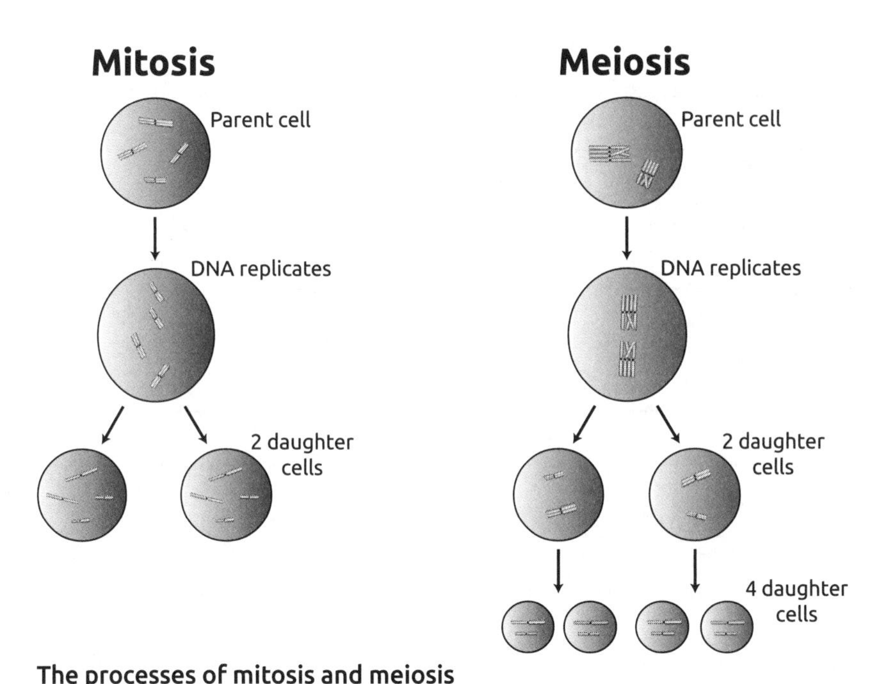

The processes of mitosis and meiosis

Helpful Hint

How can you remember which is which? Mitosis produces two twin cells, while meiosis produces egg and sperm cells.

Tissue

Tissues are made of similar cells that work together to perform a specific activity. For example, the muscle tissue in our bodies is made of millions of muscle cells. Bones, nerves, and blood are other examples of tissues in our bodies.

Organ

Organs are made of different tissues that work together to perform a specific function. The heart, brain, and liver are examples of organs in our body. For example, the heart contains muscle tissue and nerve tissue. The largest organ in the human body is skin.

Systems

Systems are a combination of organs and tissues that work together to perform a specific function. For example, several organs, such as the stomach and the intestines, work together to break down food in our digestive system. Other systems include the skeletal, muscular, and respiratory systems. There are 12 systems in our bodies.

LESSON REVIEW

Complete the activities below to check your understanding of the lesson content. The Unit 2 Answer Key is on page 152.

Vocabulary

Write definitions in your own words for each of the key terms.

1. cell cycle ______________________________

____

2. chromosome ______________________________

____

3. organelle ______________________________

____

Briefly describe the following processes.

4. mitosis ______________________________

____

5. meiosis ______________________________

____

Skills Practice

Fill in the blanks.

6. Organelles reside inside a jelly-like substance in the cell called the ____________.

7. Each human cell has ____________ chromosomes placed in a set of 23 pairs.

8. The cell division that produces new cells for growth or repair of damaged or old cells is called ____________.

9. Eggs and sperm are made through the process of cell division called ____________.

10. The largest organ in our bodies is ____________.

Answer the questions based on the content covered in the lesson.

11. The nucleus monitors all the following activities within a cell EXCEPT which one?

 A. growth of the cell

 B. reproduction of the cell

 C. metabolism within the cell

 D. allowing substances to cross through the membrane

12. Briefly describe two major roles of the nucleus.

____

____

13. Select ALL the options that could be a function of the cell.

    A. breaking down food particles

    B. attaching to other cells

    C. reproducing new cells

    D. gathering genetic material from the environment

14. Rearrange the following words in order of level of organization, beginning with the cell:

    organ, system, tissue

    cell → ______________ → ______________ → ______________

15. Name three different systems in the human body.

    ______________ ______________ ______________

16. Name three different organs in the human body.

    ______________ ______________ ______________

17. Select ALL the options that are organelles inside a cell.

    A. nucleus

    B. DNA

    C. mitochondria

    D. cell membrane

DNA AND CHROMOSOMES

Key Terms

amino acid

chromosome

codon

DNA

gene

protein

RNA

transcription

translation

You may have seen ads for home testing kits that will help you discover your ancestral origins by analyzing your genes. Doctors can use similar tests to determine if you inherited genes that indicate a high risk of certain diseases, such as breast cancer. What are genes? How are they inherited? Why are they so important?

From Chromosomes to Proteins

Humans have over 20,000 genes, contained within 23 pairs of chromosomes. A **chromosome** is a long strand of **DNA** that is curled up tightly to fit in a small space. One chromosome in each pair came from each of our parents. A **gene** is a section of the chromosome that contains the information for making a specific protein; each **protein** has a specialized function in our cells.

The proteins made according to the instructions in our genes can be used to make more cells as we grow. They are also used to send signals to other cells or work as helpers in cell functions, such as digestion and respiration.

Helpful Hint

Notice that the word "gene" is shorter than the word "chromosome." Likewise, genes are shorter strands that make up the longer strands that are chromosomes.

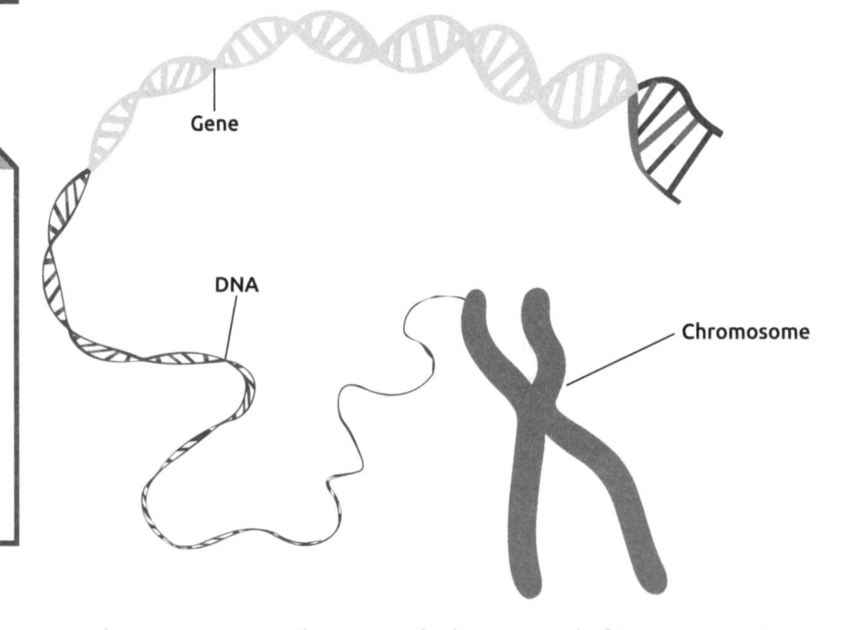

A chromosome uncoils to reveal a long strand of DNA containing many genes

Transcription and Translation

Chromosomes are kept in the nucleus of the cell for protection, since any damage to the chromosomes would cause permanent damage to the cell. Copies of the instructions on the genes are sent out instead. The instructions are held in the sequence of the building blocks making up the strand of DNA in the gene. These building blocks each contain one of four bases: adenine, thymine, guanine, and cytosine, represented by the letters A, T, G, and C.

DNA is actually a double strand of these building blocks; the bases on one strand connect with the bases on the other strand in a specific pattern. A connects only to T, and G connects only to C. Both strands carry the same code, but with different bases. For example, if one strand contains the sequence ATGC, the other strand, called the complementary strand, would have TACG in the same place.

To send a copy of a gene outside the nucleus, the cell makes a single-stranded copy of the DNA segment. The copy uses the base uracil (U) in place of the base thymine and is called messenger **RNA**, or mRNA. This process is called **transcription**. RNA is used for other functions as well, but it is always single-stranded, and it always has U instead of T.

The mRNA moves out of the nucleus so other parts of the cell can use it to make the protein for which it carries the code. Proteins are long molecules made up of many small parts called **amino acids**. Humans use 20 different amino acids in different combinations to make any protein. The cells make the proteins by connecting the amino acids according to the order given by the mRNA. This process is called **translation**.

Helpful Hint

When you copy a speech onto paper, you are transcribing it; it contains the same information in a different form. RNA contains the same information as DNA but in a different form.

Helpful Hint

When you change a sentence in English to another language, you are translating that sentence. The sequence of amino acids in a protein is a translation from the mRNA sequence.

The bases on the mRNA strand are "read" by the cell in groups of three. Each combination of three bases, called a **codon**, indicates a certain amino acid. The order of the bases in the codon is very important. For example, CAU codes for histidine (His), while UAC codes for tyrosine (Tyr). If the order of bases is changed, then the identity of the amino acids changes. This may cause the resulting proteins to not work properly.

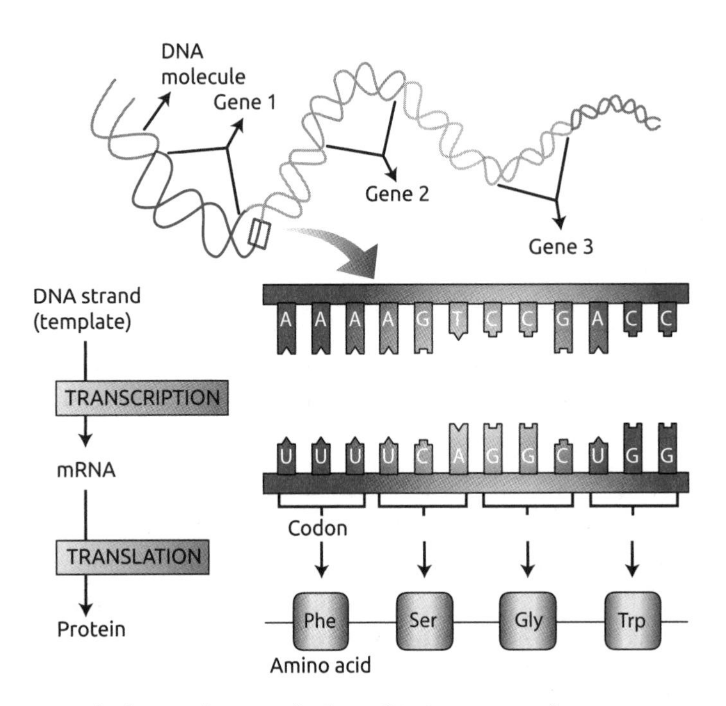

Translation and transcription of DNA to proteins

LESSON REVIEW

Complete the activities below to check your understanding of the lesson content. The Unit 2 Answer Key is on page 152.

Vocabulary

Write definitions in your own words for each of the key terms.

1. amino acid ____________________

2. chromosome ____________________

3. codon ____________________

4. DNA ____________________

5. gene ____________________

6. protein ____________________

7. RNA ____________________

8. transcription ____________________

9. translation ____________________

Skills Practice

Answer the questions based on the content covered in the lesson.

10. Write the DNA sequence for the complementary strand of AACTGATTACA. ____________________

11. Write the sequence for the RNA sequence transcribed from AACTGATTACA. ____________________

12. Which DNA sequence codes for the amino acid serine (Ser)?

    A. UCA

    B. TCA

    C. AGT

    D. TGA

MECHANICS OF INHERITANCE

Key Terms

genotype

phenotype

Punnett square

"I'm not sure where she got her blue eyes." This statement might be made about a child who has a different eye color from the rest of the family. Traits like eye color or hair color may seem to differ randomly from person to person, but they are in fact predictable. Traits occur according to certain patterns.

Genetic Inheritance

Let's start simply and investigate the genetics of a plant. A tall bean plant (represented by TT) is crossed with a short bean plant (represented by tt); the offspring of this plant are tall bean plants. The gene for tall height (T) is dominant, and the gene for short height (t) is recessive. When the two are combined (Tt), the dominant trait is the one physically displayed. The phenotype of this trait in the offspring is tall. **Phenotype** refers to the form of a trait that is visible in an organism. The **genotype** of the offspring is the actual genetic makeup. The genotype of the new bean plant would be Tt, describing the two genes for height that are inside the organism. Not all traits are inherited in this way; genetics can be quite complex.

Scientists use the **Punnett square** to help them determine the traits an offspring might have. This is a diagram that shows the possible combinations of genotypes from parents to offspring. Let's look at the bean plants mentioned previously. The gene from each parent is written in each outer square. One gene from each parent is combined in the inner squares to show the possible gene combinations in the offspring. The following Punnett square shows the possible traits of the bean plant offspring.

Height of Bean Plants (parents TT × tt)

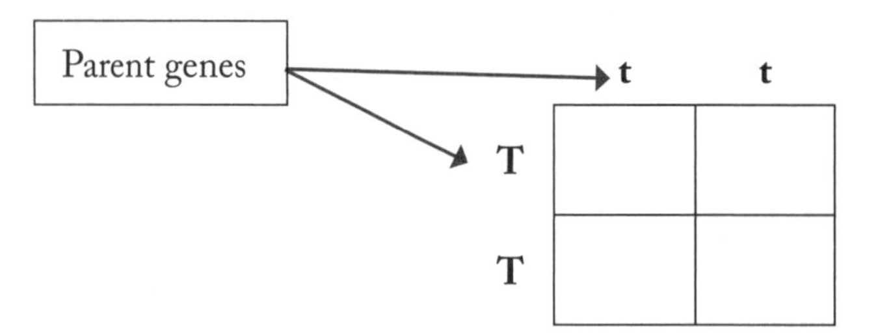

Height of Bean Plants (parents TT × tt)

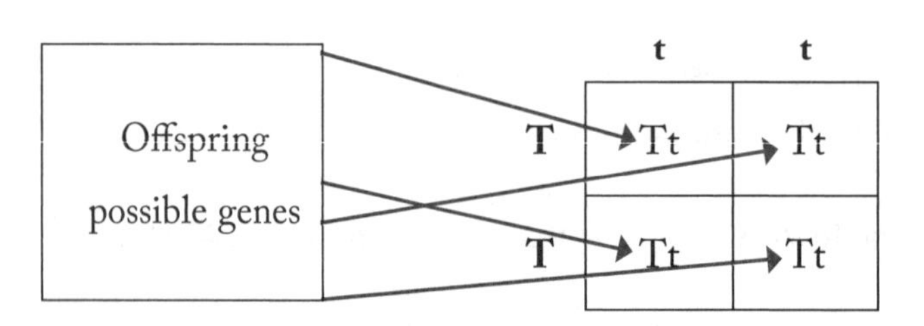

The Punnett square shows that four out of four, or 100 percent, of the offspring will be tall plants with the Tt genotype.

If the tall bean plant offspring (Tt) is crossed with a short bean plant (tt), the results will be different. There is a 50 percent chance that this offspring will be tall (Tt) because two of the four outcomes are Tt. There is a 50 percent chance it will be short (tt) because two of the four possible outcomes are tt. To have a 100 percent chance (or all four of four outcomes) that the offspring is a short bean plant, both parents would have to be short bean plants (tt), since this is a recessive characteristic.

This Punnett square shows the possible traits of the second generation of offspring.

Height of Bean Plants

	t	t
T	Tt	Tt
t	tt	tt

Helpful Hint

Genotype describes the genes for a trait that are inside an organism. Phenotype is a description of the physical trait as determined by the genes. To remember the difference, it may help to think about the first part of each word: genotype starts with a word like genes; phenotype starts with a *ph*, like the word physical.

Unit 2 Lesson 4 LESSON REVIEW

Complete the activities below to check your understanding of the lesson content. The Unit 2 Answer Key is on page 152.

Vocabulary

Write definitions in your own words for each of the key terms.

1. genotype ____________________

____

2. phenotype ____________________

____

3. Punnett square ____________________

____

Skills Practice

Use the Punnett square below to answer questions 4 and 5.

The height of the offspring species of plant is shown using the Punnett square. Tall plant height is the dominant form of the trait and is indicated by T. The recessive trait is a short plant, indicated by t.

Height of Bean Plants (parents TT × Tt)

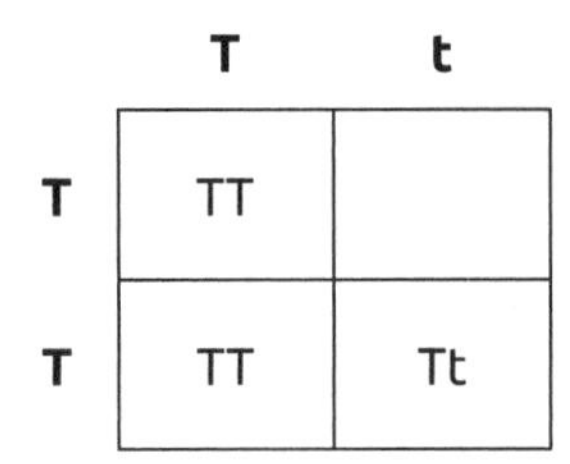

	T	t
T	TT	
T	TT	Tt

4. Which genotype should be in the blank space in the Punnett square? ______________

5. What is the chance that an offspring will be a tall plant?

   A. 25%, or 1 in 4

   B. 50%, or 2 in 4

   C. 75%, or 3 in 4

   D. 100%, or 4 in 4

CHANGES IN DNA

The English peppered moth is typically a white moth with black speckles. This helps it blend in to lichen-covered tree trunks during the daytime. Every once in a while, a black peppered moth is born. The black moths do not blend in well with their environments, and they usually do not live to pass on this trait to their offspring.

In the Industrial Revolution of the 1800s, however, the people in British cities started burning coal. As a result, soot was left on surfaces all over the cities, including trees. Then the black peppered moths blended in better than the white ones. They lived to pass on their coloring to their offspring, increasing the number of black peppered moths in cities in England.

Key Terms

allele
dominant
epigenome
gene
genome
mutation
recessive

Alleles

The color of the English peppered moth is determined by the insect's **genes**. A gene is made up of DNA, and it determines the traits of an individual. Organisms with two parents have two copies of every gene, one from each parent. Genes can have different forms, called **alleles**. The alleles determine the individual's genotype—if both alleles are the same, the individual is homozygous for that genotype. If the alleles are different, then the individual is considered heterozygous for the genotype. In the example of the bean plants from Lesson 4, TT and tt are homozygous; both alleles are the same. For the heterozygous genotype Tt, the alleles are different.

The alleles contribute to the phenotype of an organism. Let's look at the example of the English peppered moth. The allele for the dark-colored moth, A, is **dominant**, and the allele for the light-colored moth, a, is **recessive**. This means that if the A allele is present, the moth will show that phenotype for dark color. If the A allele is not present, then the moth will be light-colored.

Moth	Genotype	Phenotype
	AA Aa	Dark color
	aa	Light color

CHANGES IN DNA

Mutations

Before the Industrial Revolution, most peppered moths were light-colored. However, every so often, a **mutation**, or change, occurred in the DNA of an individual moth. This caused the dark-colored allele to appear.

A mutation can be beneficial, harmful, or neutral. In the case of the peppered moth, before the Industrial Revolution, the mutation for dark color was harmful. The dark moths would stand out on the white trees and be eaten by predators. As more soot was deposited on trees from burning coal, the dark-colored mutation became beneficial. Then the dark-colored moths blended in, and the light-colored moths stood out.

Mutations occur quite frequently, but most are neutral. When a cell copies its DNA before replicating, an error can occur. The error is similar to someone copying a paragraph in a document and making a typo in one of the sentences. The most common type of mutation is a substitution, where one single base is substituted for another:

CCATG: This is the original sequence.

CCATA: This is the sequence with a substitution—the A is substituted for the G.

Some other types of mutations are deletions and insertions. In a deletion, a section of bases is completely deleted from the sequence:

CCATGAGTC: This is the original sequence.

CC~~ATG~~AGTC: A section is deleted.

CCAGTC: This is the sequence with the deletion.

In an insertion, extra base pairs are inserted into the DNA:

CCATGG: This is the original sequence.

CCAGTACTGG: The bases GTA were added into the sequence.

Real-World Connection

Cancer is caused by genes that have undergone mutation. The mutated cells then continue to replicate, causing a tumor to form. Abnormal methylation is found in many cancers, which would interfere with the epigenome switching genes on or off. The link between the epigenome and cancer is not clearly understood, but the evidence showing the association is growing.

On/Off Switches

Every organism has a **genome**—its entire set of DNA and the genes inside it. Covering the genome are chemical tags, called the **epigenome**. The epigenome wraps around the genome and acts as a chemical "switch." It keeps inactive genes from being expressed and lets active genes be recognized. DNA is in every cell of our bodies, but not every gene is needed in every cell. For example, the same genes would not be expressed in both a skin cell and an eye cell.

The epigenome can be influenced by an individual's environment and life experiences. Diet, stress, and activity can give signals to the epigenome, which then carries a protein to a specific sequence of DNA. It then acts like a switch, either turning genes on or shutting them off.

An important classification of epigenetic tags is methyl groups. Methyl groups are one type of molecule responsible for turning off gene expression. If the pathway that makes these groups is altered, then gene expression can be changed. Animal studies have shown that a pregnant mother's diet can affect methylation in the offspring for life.

LESSON REVIEW

Complete the activities below to check your understanding of the lesson content. *The Unit 2 Answer Key is on page 152.*

Skills Practice

Answer the questions based on the content covered in the lesson.

1. A sequence of bases in a gene is CCATTG. After the DNA is copied, the base reads CCAGGTTG. Which type of mutation has occurred?

   A. completion

   B. deletion

   C. insertion

   D. substitution

2. How do methyl groups influence genes?

   A. They cause the epigenome to mutate and change genes.

   B. They are responsible for ordering the base pairs in DNA.

   C. They cause the nucleotides in DNA to become recessive.

   D. They are responsible for shutting down the expression of genes.

3. A recessive phenotype will be expressed if ________________________ alleles are recessive.

EVOLUTION

Key Terms

adaptation

common ancestor

natural selection

speciation

variation

In 1982, a new species of bird, the large ground finch, appeared on the island of Daphne in the Galapagos Islands. The medium ground finch had already been living on the island. Now it had a direct competitor for seeds on the ground.

Both birds liked the large seeds on the ground, but the larger-beaked large ground finch had an easier time cracking them open. The medium ground finch couldn't compete with the new bird that was more than twice its size.

Since the arrival of the large ground finch, the medium ground finch has evolved. It now has a smaller beak, which allows the medium ground finch to eat the smaller seeds on the ground, which the large ground finch has no interest in. The medium ground finch is able to survive in its environment with a smaller beak.

The medium ground finch (top) has a smaller beak than the large ground finch (bottom).

A Common Ancestor

During his voyage on the HMS *Beagle* from 1831 to 1836, Charles Darwin took notice of the species on the Galapagos Islands. He remarked that it seemed as if all the species of finches seemed to have come from one common bird, and then were modified for different purposes. Although he didn't know it at the time, he was correct.

Charles Darwin noted the different beak shapes of the finches on the Galapagos Islands. Each shape is adapted to eat a different type of food.

(Source: Charles Darwin, "The Voyage of the Beagle," 1845)

The Galapagos Islands is home to 14 species of finches. All 14 of these species came from a **common ancestor** from Central or South America. It came to the Galapagos Islands 2 or 3 million years ago. This ancestral species then branched out, becoming the 14 species that are known today.

On a broader scale, all life on earth has evolved from a common ancestor. About 3.8 million years ago, the first life appeared on earth as a single-celled organism. From that single-celled organism, life has evolved into all the species that occupy every niche on earth. The bacteria living in thermal vents, the flowers that grow in your garden, the cat that curls up with you at night—all came from the same first species.

Natural Selection

Each species of Galapagos finch has a different, specialized beak, allowing it to eat different foods and access food in different ways. The warbler finch has a thin, sharp beak, which it uses to spear insects. The ground finch has a blunt beak that allows it to crack open seeds, the woodpecker finch can drill holes in trees, and the vampire finch drinks other birds' blood. If they all came from the same ancestor, how did they become so different?

Let's look at the example of the medium ground finch and the large ground finch. The medium ground finches experienced competition for their food source. If the birds couldn't eat, then they would die. But some birds in the population had smaller beaks and could crack open the smaller seeds, and these birds survived to pass on their traits to the next generation. This process, in which the individuals with beneficial traits survive, is called **natural selection**.

Adapt or Die

If the medium ground finch had kept a larger beak, the species would have died out. The smaller beak was an adaptation that allowed for the continuation of the species.

An **adaptation** is a trait that is found in most members of a species. It exists because it allows the species to be better able to survive in its environment. There are many types of adaptations. The beak of the medium ground finch is an example of an adaptation of an anatomical feature. Adaptations can also be behaviors, such as being able to better evade a predator. An adaptation can be a chemical in the body that allows an animal to survive at higher temperatures. Another example of an adaptation may allow an insect to blend in with its environment.

Natural selection is a result of **variation** within species. When reproduction occurs, different sets of genes combine. Sometimes there are mutations. These can lead to small differences in individuals. Over time, however, these differences can become adaptations.

New Species

A species is generally defined as a group of individuals that can produce offspring in nature. **Speciation** occurs when a group becomes two or more different species.

Speciation begins when groups of a species stop mating with each other. Sometimes there is a geographic barrier. Other times they move far away from each other. The groups adapt to their different environments. Eventually, they become genetically different and can no longer interbreed. They have become different species.

LESSON REVIEW

Complete the activities below to check your understanding of the lesson content. The Unit 2 Answer Key is on page 152.

Vocabulary

Write definitions in your own words for each of the key terms.

1. adaptation ______________________________

____

2. common ancestor ______________________________

____

3. natural selection ______________________________

____

4. speciation ______________________________

____

5. variation ______________________________

____

Skills Practice

Answer the questions based on the content covered in the lesson.

6. How does natural selection lead to an adaptation?

____

____

7. A species lives in a forest habitat. The trees in the habitat become taller, making the leaves more difficult to reach. The individuals with longer necks can reach the leaves, while the individuals with shorter necks cannot. Which best describes the likely future of the species?

   A. Speciation will occur, creating one species with short necks and one species with long necks.

   B. Variation will occur, allowing the individuals with short necks to be able to grow longer necks.

   C. Natural selection will occur, and the species will consist of a majority of individuals with long necks.

   D. Common ancestry will occur, with all short-necked and long-necked individuals tracing their lineage back to one individual.

LESSON REVIEW

8. Every species on Earth comes from the same _________________________.

9. Two groups of insects of the same species become separated after a storm. A scientist notices that over time, they became two different species. How does the scientist know that they are no longer the same species?

   A. They are different colors.

   B. One group cannot fly.

   C. One group is larger than the other.

   D. They cannot create offspring with each other.

An Overview of the Human Body Systems

You probably know about many of the organs of your body, such as the brain, heart, and lungs. But do you know how these organs work with other organs to keep your body working and to allow you to run, eat, breathe, heal, and think?

Key Terms

homeostasis

hormone

negative feedback

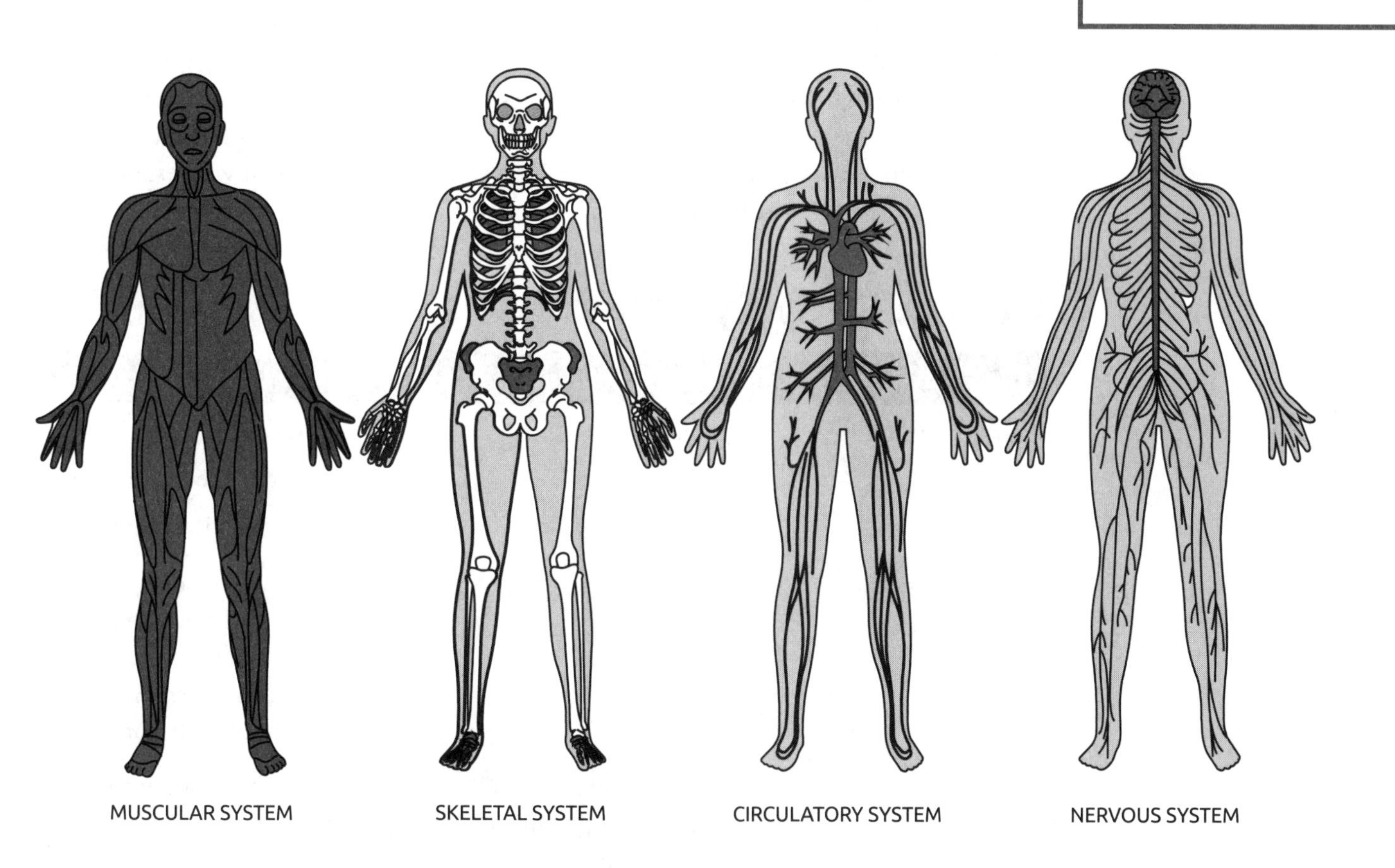

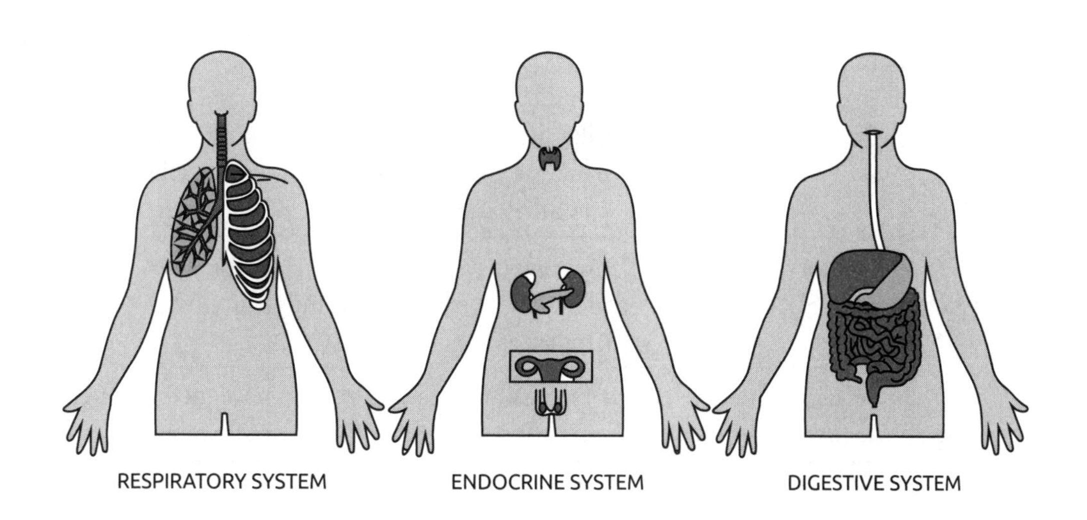

How Do the Systems Work Together?

As you can see from the chart, many systems depend on other systems to help do their jobs. For example, the muscular system moves the bones of the skeletal system, but it also moves food through the digestive tract. As the food is pushed through the digestive system, it is broken down into nutrients. The nutrients are used for many functions, such as providing the body with energy and helping to build and repair tissues. The glands of the endocrine system produce **hormones**. Hormones are like chemical messengers. They travel to different organs to control activities like growth and the absorption of nutrients. The circulatory system is responsible for moving all these nutrients, as well as gases, hormones, and other important substances, through the body. The nervous system is like the conductor, responding to conditions in and out of the body. For instance, if the nervous system senses that your hand is touching something very hot, it will respond by instructing the muscles of the hand to move it away.

Human Body Systems

System	Components	Purposes
Skeletal system	Bones	Support for the body Attachment site for muscles Protection of vital organs
Muscular system	Muscles	Movement, including the beating of the heart, movement of food through the digestive system, and movement of body parts
Digestive system	Mouth, stomach, small intestine, large intestine, and other organs	Breakdown of food into nutrients
Respiratory system	Mouth and nose, trachea, bronchi, lungs	Gas exchange, or movement of oxygen into the body and carbon dioxide out of the body
Endocrine system	Glands, including the pancreas, pituitary, thyroid, and adrenal glands	Regulation of growth, metabolism, and reproduction
Circulatory system	Heart and blood vessels, including veins, arteries, and capillaries	Transport of materials around the body
Nervous system	Brain and nerves	Regulation of conditions in the body

Regulation of the Workings of the Organs

Homeostasis is a term that describes the way in which the body keeps the conditions in the body at an ideal state for the functioning of all the organs. For instance, there is a certain temperature (98.6°F) at which the body works best, as well as an optimal amount of oxygen and sugar in your body. If these levels are off, the body takes action to get them back to normal.

In order to regulate the conditions, the body must gather information and respond to it. This usually involves several systems working together. Most commonly, the body uses negative feedback. **Negative feedback** can be described as a reaction to a change in a way that brings it back to its original state. For instance, you may know that much of the calcium in our bodies is found in the bones and teeth. We also need a small amount of calcium in our bloodstream. If that amount gets too large and there is too much calcium in the blood, the thyroid gland (part of the endocrine system) releases a hormone called calcitonin. This hormone causes calcium to be taken up by the bones, decreasing the amount in the bloodstream. Another example of homeostasis involves the respiratory, circulatory, and nervous systems. Remember that carbon dioxide is a waste product of respiration. If there is a high level of carbon dioxide in your bloodstream, your brain tells your lungs to breathe faster. You exhale more, releasing more carbon dioxide. The endocrine system also plays a big role in maintaining a stable environment within the body. For example, if the levels of sugar in the blood are high, a gland called the pancreas releases insulin into the bloodstream. Insulin helps the body's cells absorb sugar from the blood.

Helpful Hint

If you break down the word *homeostasis* into two parts—*homeo-* and *stasis*—you get a clue to its meaning. The prefix *homeo-* comes from the Greek word *homoios,* which means "similar" or "alike." *Stasis* is a condition in which things are in a state of balance. So, in homeostasis, the body seeks to keep the same balance of the conditions in the body so it works at its peak performance level.

Unit 2 Lesson 7 LESSON REVIEW

Complete the activities below to check your understanding of the lesson content. The Unit 2 Answer Key is on page 153.

Skills Practice

Fill in the blanks.

1. The ______________________ system is responsible for breaking food down into useable nutrients.
2. The ______________________ system allows oxygen to enter the bloodstream.
3. The ______________________ system works with the skeletal system to allow you to swim.
4. Without the ______________________ system, oxygen could not reach the brain.
5. The body uses negative feedback to maintain ______________________.

LESSON REVIEW

6. An example of homeostasis is sweating when you are hot. Sweating helps maintain your body's ________.
   - **A.** sugar level
   - **B.** oxygen level
   - **C.** calcium level
   - **D.** temperature level

7. Which system protects your brain and lungs?
   - **A.** endocrine
   - **B.** muscular
   - **C.** nervous
   - **D.** skeletal

8. Which system works with the digestive system to allow nutrients to reach the muscles?
   - **A.** circulatory
   - **B.** muscular
   - **C.** nervous
   - **D.** skeletal

9. Which system helps move food through the small intestine?
   - **A.** circulatory
   - **B.** muscular
   - **C.** nervous
   - **D.** skeletal

10. Which system releases hormones that help regulate growth?
    - **A.** circulatory
    - **B.** endocrine
    - **C.** nervous
    - **D.** skeletal

11. Which system works with the circulatory system to allow oxygen to reach the cells?
    - **A.** circulatory
    - **B.** muscular
    - **C.** respiratory
    - **D.** skeletal

12. Which system uses the senses to receive information about the environment and then acts on it to maintain homeostasis?
    - **A.** circulatory
    - **B.** nervous
    - **C.** respiratory
    - **D.** skeletal

13. Which system works with the nervous system to control homeostasis?
    - **A.** endocrine
    - **B.** muscular
    - **C.** respiratory
    - **D.** skeletal

DISEASE AND PATHOGENS

If you have ever had a cold or sore throat, you have had a bacterial infection or virus. Bacteria and viruses cause most illnesses, including the common cold, strep throat, and the stomach flu.

Key Terms

antibody

antigen

bacteria

host

immunization

infection

pathogen

toxin

vaccine

virus

Bacteria and Viruses

An organism that causes a disease is called a **pathogen**. When there is a pathogen in the body, it is called an **infection**. **Bacteria** are one-celled organisms that can be seen only through a microscope. Bacteria have DNA and reproduce quickly in the body. Many bacteria give off **toxins** (poisons) that damage your cells and tissues. Bacterial infections include tuberculosis, strep throat, and cholera. Generally, specific illnesses are caused by specific bacteria; for instance, tuberculosis is caused by bacteria called *Mycobacterium tuberculosis.* Bacterial infections can often be cured using antibiotic medications. Antibiotics kill bacteria by targeting their growth and reproduction.

Viruses are even smaller than bacteria. Unlike bacteria, viruses have either DNA or RNA and can reproduce only if they are inside the cell of a living thing. Viruses are very difficult to kill. Antibiotics do not work on viruses because viruses have a different structure and reproduce in a different way from bacteria. In fact, taking an antibiotic when you have a viral infection may kill the good bacteria inside your body. Viral infections include HIV, influenza (the flu), chickenpox, and measles.

The image belows shows the basic structure of bacteria.

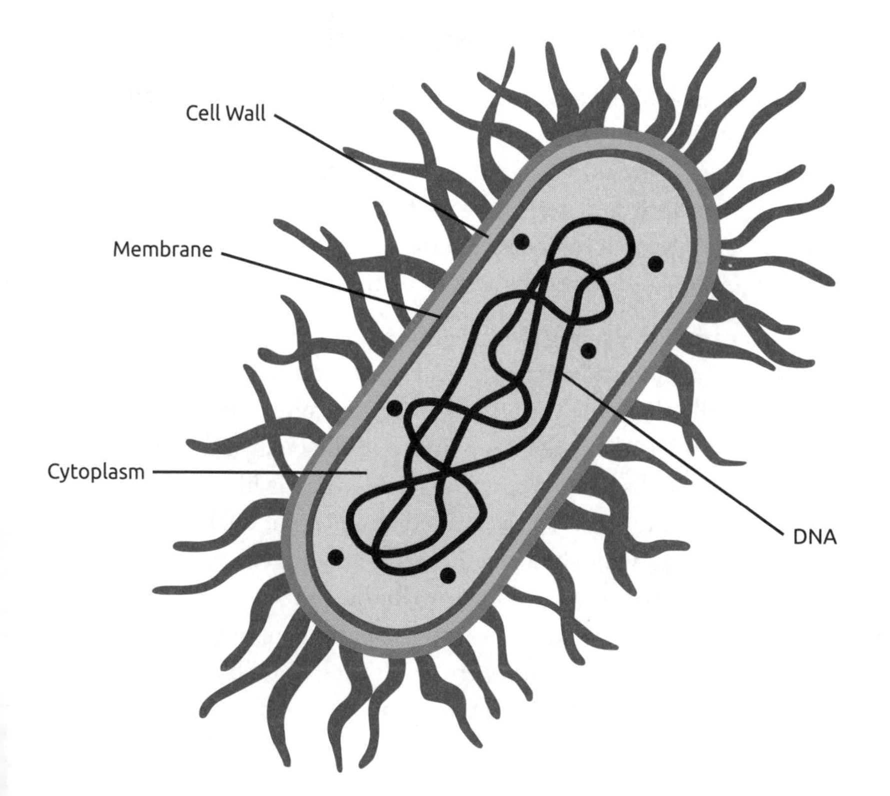

Viruses come in a variety of shapes, but they always have the same basic structure of a protein coating surrounding genetic material. The image below is of HIV, the virus that causes AIDS.

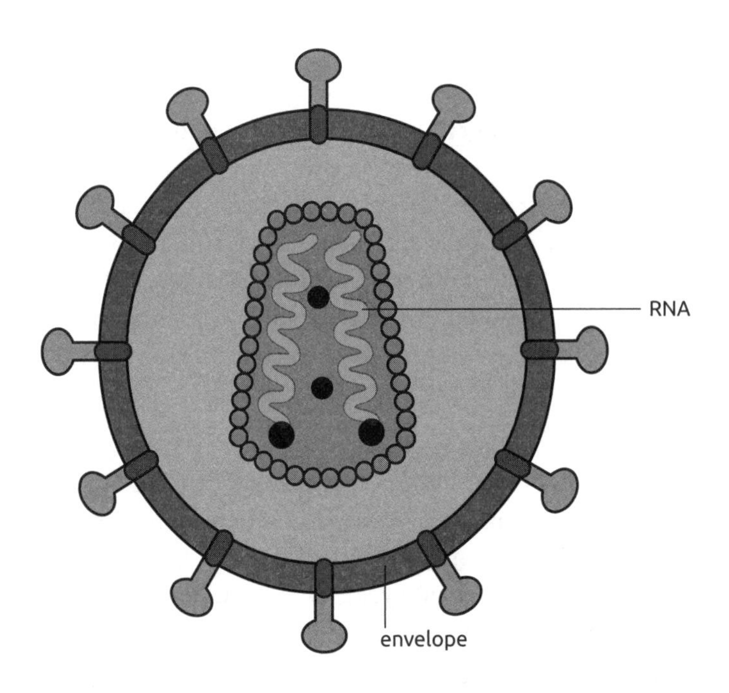

How Do Bacteria and Viruses Make Us Sick?

In order to cause disease, a pathogen must get into the body. Pathogens are spread in a number of ways, including touching something that has germs on it and having direct contact with someone who is sick. The skin acts as the first defense against pathogens, but if you have a cut in your skin, bacteria and viruses can get inside. They can also enter the body through the nose or mouth, so you can become infected if you share food or water with an infected person, or are in close contact when a sick person coughs or sneezes.

Once inside a **host**—which can be any living thing—different pathogens affect the body in different ways. For instance, a virus will enter the host's cells and make them create more viruses, destroying the cells along the way. Bacteria can reproduce so quickly that they cause problems even in cells that have not been infected; they may also release toxins that kill cells. To fight infection, we have an immune system, which is like the body's own army. All pathogens have a protein on their surface called an **antigen**. Each pathogen has its own unique antigen. When a white blood cell detects an antigen that is foreign to the body, it signals the immune system to produce **antibodies** that are unique to that antigen. The antibody binds to the antigen and destroys the pathogen. Some white blood cells also destroy any cells infected with the pathogen. Still other white blood cells release chemicals that work against the toxins that the bacteria produce.

Real-World Connection

Not all bacteria are bad—in fact, it is believed that only 1% of the planet's bacteria cause disease in humans. Most bacteria are helpful. For example, bacteria help break down food in your digestive system and also release vitamins.

How Can You Prevent Infection?

The number one way to prevent an infection is to wash your hands. It is best to wash your hands with soap and water for 20 seconds. You should also avoid touching areas not covered by skin, such as your eyes, nose, and mouth. Another way to prevent catching an illness is to make sure you receive all your **immunizations**. When you are immunized, you receive a vaccine. A **vaccine** is a dead or inactive pathogen. Your body detects the antigen and produces antibodies for it; even though you do not become ill, the cells that make the antibody remain. If you become exposed to that pathogen again, your immune system can make the antibodies for it much faster. You are immune to that pathogen because it will not make you sick. You can also become immune to many diseases after you have been infected with them once. Since some of those diseases can cause permanent damage to your health, getting immunity with a vaccine is safer for most people.

Unit 2 Lesson 8 LESSON REVIEW

Complete the activities below to check your understanding of the lesson content. *The Unit 2 Answer Key is on page 153.*

Vocabulary

Fill in the blanks.

1. A ________________________ causes disease.

2. A ________________________ is a substance that some bacteria release that can damage the cells.

3. A ________________________ can reproduce only inside a host's cell.

4. ________________________ are single-celled organisms that can reproduce very quickly in our bodies.

5. When bacteria is disrupting your body, you have a(n) ________________________.

Skills Practice

Select the correct option.

6. The (skin, white blood cell) is the first line of defense against infection.

7. The (immune, nervous) system defends your body against disease.

8. Antibiotics work to rid the body of (bacteria, viruses).

Answer the questions based on the content covered in the lesson.

9. Which of the following are good practices to prevent diseases from spreading?
   A. not sharing drinks
   B. proper hand washing
   C. covering your mouth when you sneeze
   D. all of the above

10. What causes you to feel feverish when you have the flu?
   A. the immune system fighting the virus
   B. the virus reproducing in the cells
   C. the bacteria producing toxins
   D. all of the above

NUTRIENTS AND ENERGY

Key Terms

amino acid

calorie

fermentation

minerals

photosynthesis

vitamin

Humans are one of the few species to eat a wide variety of plant-based and animal-based food. The food that we eat breaks down in our bodies to provide us with energy as well as vitamins and minerals that we need to stay healthy and function throughout the day. Even though we may eat both plants and animals, the animals we eat depend on plants for their food, so all of our food ultimately comes from plants. Where do plants get their food?

Photosynthesis

Photosynthesis is a process in plants that uses sunlight to produce sugar and starch. The leaves of a plant use the energy from light to convert water and carbon dioxide into sugar and oxygen. The plants then use the sugar to make cellulose, their main building material. They also store the sugar by converting it to starch. Photosynthesis takes place in chloroplasts. Chloroplasts are organelles within plants cells that are filled with chlorophyll. Chlorophyll is the green molecule that gives plants their color.

Respiration is the process of taking in oxygen and releasing carbon dioxide to produce energy. Respiration takes place in the cells of both plants and animals. In cellular respiration, the chemical bonds in sugar and oxygen are broken down. It is through breaking these bonds and forming new bonds to produce carbon dioxide and water that energy is released to be used by the cell.

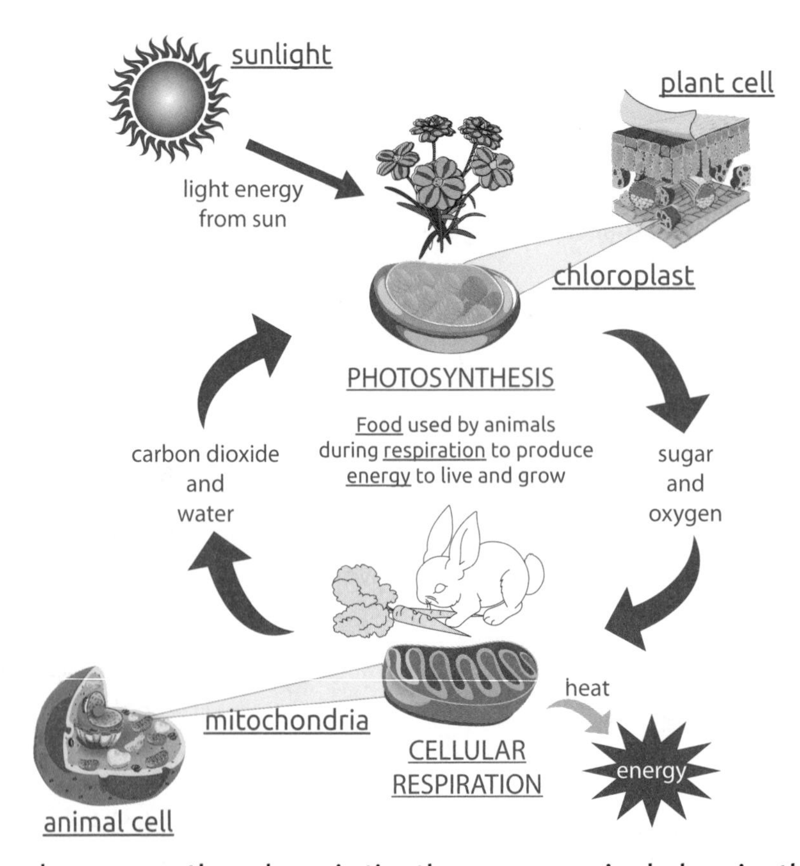

Plants release energy through respiration the same way animals do, using the sugar they produced during photosynthesis.

Fermentation

Fermentation is the process of producing energy when there is not enough oxygen to break down the sugar. In cellular respiration using oxygen, energy is produced by breaking down sugar into carbon dioxide and water. In fermentation, sugar is also broken down to produce energy. But without oxygen, carbon dioxide and lactic acid are the products. This is why fermented food tastes sour. Another example of fermentation is the burning we feel in our muscles during heavy exercise. When our heart and lungs cannot provide enough oxygen during exercise, our body makes energy through fermentation, and lactic acid is produced in our muscles. Other forms of fermentation can produce alcohol instead of lactic acid as a byproduct.

Calories

A **calorie** is a measurement of energy. Food calories measure the amount of energy available for use in food. Scientists measure the amount of calories in food by burning a small amount of it and calculating the amount of energy produced. A similar process takes place in our bodies when we use respiration to "burn" food and gain energy.

An average human needs about 2,000 calories a day. The calories we take in through eating food give us energy for activities and movement. If we consume more calories than we use, our cells store the excess energy in the form of fat for later use. Similarly, other living organisms need calories to function properly.

Amino Acids

Amino acids are the building blocks of proteins in our bodies. About 20 percent of the human body is made of protein. Proteins are essential for many biological processes in our bodies. There are 20 amino acids that our bodies need to live and grow, but we can produce only 11 of them. The remaining 9 are called the essential amino acids because we need to consume them through food to remain healthy.

Real-World Connection

Some organisms receive all their energy through fermentation, like yeast. When making bread, yeast and flour are combined. Flour is a carbohydrate, and carbohydrates are simply long sugar molecules. When yeast is added to flour, it uses the sugar in flour to produce energy. As a result, carbon dioxide is created, which causes the bread to bubble and rise. You may sometimes taste the sourness from lactic acid in bread.

Vitamins

Vitamins are organic substances needed in trace amounts for growth and health. There are 13 essential vitamins that are not produced by the human body. Vitamins work by helping reactions occur, such as the ones to build proteins and other molecules. Without the right vitamins, these reactions would occur more slowly or not at all. For example, vitamin C helps produce the proteins that keep our gums and skin healthy, while vitamin A helps produce the light-sensitive molecules in our eyes.

There are two types of vitamins: fat-soluble vitamins and water-soluble vitamins. Fat-soluble vitamins require the help of protein molecules to enter the bloodstream and can be stored in fat cells. Vitamins D, E, A, and K are examples of fat-soluble vitamins. Our bodies can make small amounts of vitamins D and K, but we need to supplement this amount through food. Vitamins C and B are examples of water-soluble vitamins. Water-soluble vitamins easily enter the bloodstream but also quickly leave the body. This is why it is important to replenish your supply of water-soluble vitamins on a daily basis.

Minerals

Minerals are inorganic substances needed in trace amounts for growth and health. Minerals are vital for development of strong bones and teeth, muscle and nerve function, and production of red blood cells. Minerals are not produced in our bodies, so we need to consume them on a regular basis. Minerals dissolve in water, and any extra amount is eliminated through urination. Calcium is an example of an essential mineral. Calcium is one of the most abundant minerals in bones and teeth. It is also a part of the signaling system in our nerves and muscles. Foods such as milk and leafy greens are rich in calcium.

LESSON REVIEW

Complete the activities below to check your understanding of the lesson content. The Unit 2 Answer Key is on page 153.

Vocabulary

Write definitions in your own words for each of the key terms.

1. amino acid ______________________________
____
2. calorie ______________________________
____
3. fermentation ______________________________
____
4. photosynthesis ______________________________
____
5. respiration ______________________________
____

Skills Practice

Select the correct option.

6. Minerals are (inorganic, organic) substances needed in trace amounts for growth and health.
7. Vitamins D and A are examples of (water-soluble, fat-soluble) vitamins.
8. Our bodies can produce only (9, 11) out of the 20 amino acids.

Answer the questions based on the content covered in the lesson.

9. All the following are examples of fermentation EXCEPT which one?
   A. making pickles
   B. making yogurt with yeast
   C. mixing lemon juice and water
   D. feeling muscles "burn" during exercise

10. Which option describes the process of cellular respiration?
   A. sunlight + carbon dioxide + water → sugar + oxygen
   B. sugar → carbon dioxide + lactic acid
   C. oxygen + sugar → carbon dioxide + water
   D. sugar → carbon dioxide + alcohol

FLOW OF ENERGY AND MATTER IN ECOSYSTEMS

Key Terms

consumer
decomposer
ecosystem
energy pyramid
food chain
food web
producer
trophic structure

All living organisms require energy to survive. For some, like plants, their energy may be in the form of sugar made from carbon dioxide in the air. For some animals, such as sheep, plants are the source of nutrients and energy. Other animals, such as wolves, eat animals in order to survive. A few species, such as humans and bears, are able to receive nutrients and energy from multiple plant- and animal-based sources. All the organisms living in an area form a complex system in which they depend on each other for energy.

Ecosystem

An **ecosystem** is a community of living organisms and nonliving things that work together to make a balanced system. A lake is an example of a natural ecosystem. Water, fish, algae, air, and sunlight are some parts of a lake's ecosystem. The organisms in an ecosystem may depend on each other and on the nonliving things for survival. If one part of an ecosystem is out of balance, the relationship between its members may change. The balance of an ecosystem may be disturbed by natural disasters or by human intervention. A healthy ecosystem has a great variety of species. A fish tank is an example of an artificial ecosystem that you may create at home.

Trophic Structures

Trophic structures are the feeding relationships between the organisms in an ecosystem. Every trophic structure is composed of trophic levels. Producers, consumers, and decomposers are on the trophic levels. **Producers** are organisms that capture the energy of the sun and make carbohydrates, or sugar. Plants are examples of producers. They store the carbohydrates in their leaves, stem, and roots. These parts of the plants are then eaten by **consumers**, such as rabbits. Other consumers will then eat the rabbit. Once the plants and animals die, decomposers will break down the parts and turn them into matter that can be used by producers. **Decomposers** are plants, animals, and fungi that feed off of the dead bodies of a living organism.

An **energy pyramid** is a model of the energy flow through an ecosystem. The shape of the pyramid shows how the amount of useful energy changes as it moves through an ecosystem's trophic levels. Scientists have calculated that about 90% of the energy at each trophic level is lost as it passes to the next level. The trophic levels in a typical energy pyramid include producers and primary, secondary, and tertiary consumers. Primary consumers are herbivores that eat plants. Some familiar examples of primary consumers are squirrels, elephants, rabbits, and goats. Secondary consumers eat primary consumers and can include carnivores such as dogs, eagles, and snakes. Tertiary consumers are sometimes called apex consumers. Examples of tertiary consumers are humans, bears, and tigers. Secondary and tertiary consumers might eat plants as well as animals.

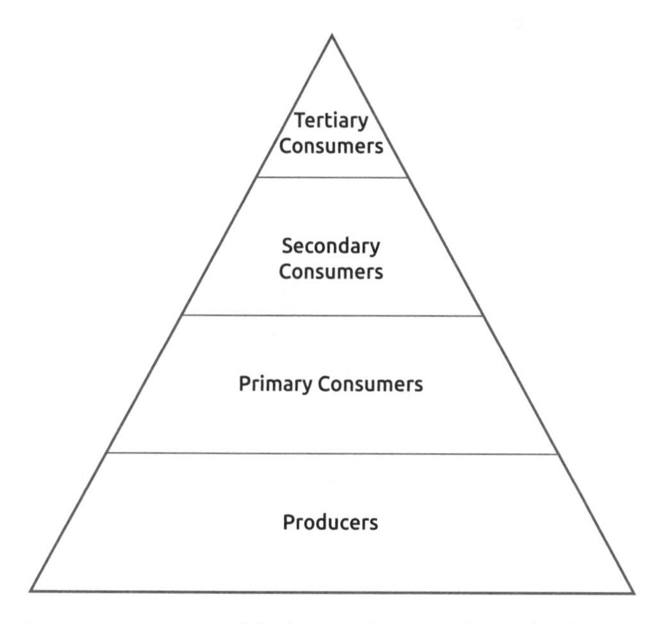

In an energy pyramid, the most energy is at the bottom, and energy is lost in each level moving toward the top.

Real-World Connection

In an energy pyramid, scientists measure the energy lost at each trophic level. They calculate the energy by estimating the biomass of a population. First, the dry weight of an average member of a species is measured. Since all living organisms are made of similar organic molecules, the dry weight is a good estimation of the stored energy. Dry weight multiplied by the population of the species gives us the biomass of the species. In an energy pyramid, producers have the largest biomass, while the biomass of consumers decreases at each level.

FLOW OF ENERGY AND MATTER IN ECOSYSTEMS

Food Web

A **food web** is a model of feeding connections among the species in an ecosystem. A food web also displays the flow of energy from one species to another. There are many trophic levels in a food web. Omnivores create multiple paths within a food web by consuming food from various trophic levels. Each linear path within a food web is a **food chain**. For example, consider the food web shown below. The connections between seeds, mice, and a hawk make up a food chain within this food web.

Forest Food Web

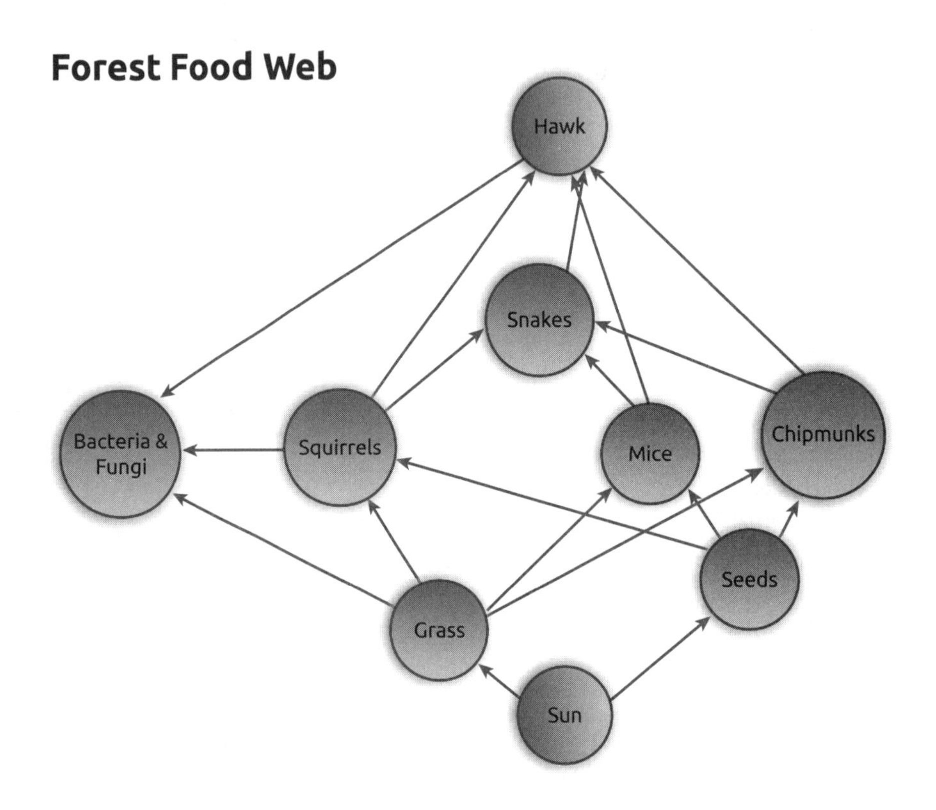

LESSON REVIEW

Complete the activities below to check your understanding of the lesson content. The Unit 2 Answer Key is on page 153.

Vocabulary

Write definitions in your own words for each of the key terms.

1. ecosystem ______________________________

____

2. food web ______________________________

____

3. trophic structures ______________________________

____

Skills Practice

Answer the questions based on the content covered in the lesson.

4. Name two living and two nonliving members of a forest ecosystem. ______________________________

____

5. What do producers do in an ecosystem and what is an example of a producer? ______________________________

____

6. What is an energy pyramid and why is it shaped like a pyramid? ______________________________

____

7. All the following are true about humans except which one?

   A. Humans are omnivores.

   B. Humans are apex consumers.

   C. Humans can use sunlight to produce sugar in their bodies.

   D. Humans' activity threatens the health of many ecosystems.

Select the correct option.

8. (Consumers, Decomposers) break down the dead body of a living organism and turn it into matter that can be used by producers.

9. (Secondary, Tertiary) consumers are also called apex consumers.

10. Each linear path within a food web is a (food chain, energy pyramid).

LESSON REVIEW

Consider the following food web in order to answer questions 11–14.

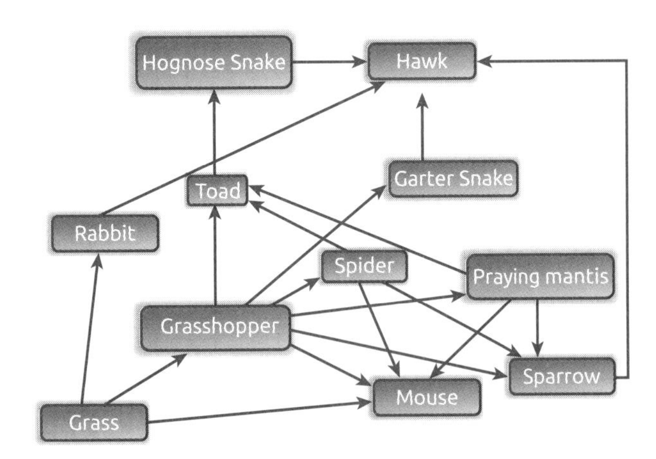

11. Identify a producer. ____________________

12. Identify a primary consumer. ____________________

13. Identify a secondary consumer. ____________________

14. Identify an apex consumer. ____________________

15. Organize the following organisms on the energy pyramid below:

    sparrow, wheat grains, owl, small snake

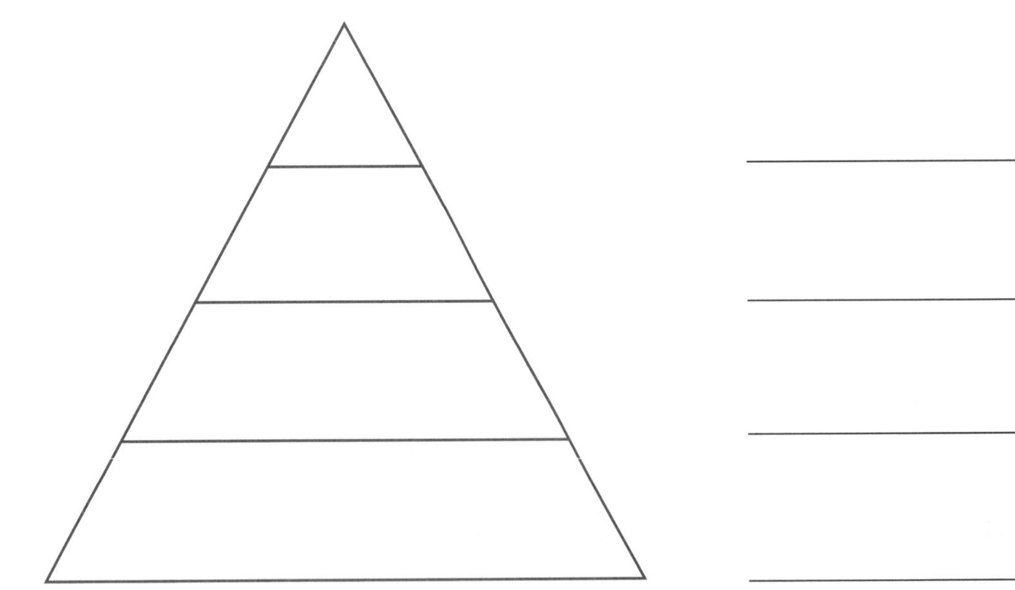

Unit 2 Lesson 11 CARRYING CAPACITY AND DISRUPTIONS OF ECOSYSTEMS

Have you ever had relatives or friends come to visit? The addition of these people can make a very noticeable difference in a household. In order for everyone to get showered, a schedule has to be made. Additional food must be purchased and prepared, and you need to find space for your guests to sleep. Sharing resources such as food and shelter can be demanding, and even one extra person can affect many things. This occurs in nature as well. Changes in populations can affect the balance of an entire ecosystem.

Key Terms

carrying capacity

invasive species

limiting factors

Carrying Capacity and Limiting Factors

Food webs show how all living things in an ecosystem are connected. The balance of nature is delicate. It can be affected by changes in populations due to natural causes, human actions, or habitat destruction. Each part of a food web is directly or indirectly connected to all other parts. A change in the population of one species can affect many other species within the food web.

Carrying capacity is the maximum population size an ecosystem can support without losing resources. Imagine if the number of rabbits in a forest increased from 200 to 300. The rabbits would eat more grass and clover. This would leave less grass for deer to eat. The deer population might fall, and so might the population of animals that eat deer. In this way, the increase of the rabbit population could disrupt the entire ecosystem. The rabbit population would have become larger than the ecosystem's carrying capacity.

The carrying capacity of an ecosystem is determined by the system's **limiting factors**. Limiting factors are things like food, water, and shelter. There are only so many of these things for organisms to share. Other animals that are competitors or predators can be limiting factors as well. These factors limit the size of a population. During spring the rabbit population may grow from 200 to 300 rabbits. There may be only enough food for 200 rabbits, so food is a limiting factor. Because of this limiting factor, the rabbits are over their carrying capacity and the population will likely fall back to 200.

CARRYING CAPACITY AND DISRUPTIONS OF ECOSYSTEMS

Real-World Connection

Earth's carrying capacity for humans has increased since prehistoric times. This is because humans have the unique ability to alter carrying capacity. We did this by developing the use of tools to hunt and to grow and collect food. The result was more plentiful food. Humans have also built shelters in places that had no natural shelter.

Ecosystem Disruption

Ecosystems can be disrupted by anything that affects one or more populations. Disruption can also be caused by changing the availability of limiting factors. Natural disasters, human activity, and invasive species are things that can disrupt an ecosystem.

Invasive species are organisms that are not native to an area. The introduction of an invasive species often has a harmful effect. An example of an invasive species is the zebra mussel. Zebra mussels are small marine animals that attach themselves to solid surfaces. They may attach to boats, rocks, aquatic plants, and animals. They were carried to the United States unknowingly on ships from Europe. They can clog pipes and damage structures. They reproduce quickly and cause competition for resources.

The structure shown below has been overtaken by invasive zebra mussels.

Humans are also a threat to ecosystems. Many habitats are destroyed when people clear land to build homes, roads, and businesses. The animals that live in these habitats lose their food sources and shelter and must find new places to live.

LESSON REVIEW

Complete the activities below to check your understanding of the lesson content. The Unit 2 Answer Key is on page 153.

Vocabulary

Write definitions in your own words for each of the key terms.

1. carrying capacity ______________________________

____

2. invasive species ______________________________

____

3. limiting factor ______________________________

____

Skills Practice

Answer the questions based on the content covered in the lesson.

4. Which would have the most significant effect on the population of squirrels in a suburban area?

   A. a change in season from fall to winter

   B. a homeowner cutting down two trees in his yard

   C. an increase in the population of the squirrel's predators

   D. a thunderstorm that breaks several branches from trees

5. Name two limiting factors for the mouse population shown in the food chain below.

____

____

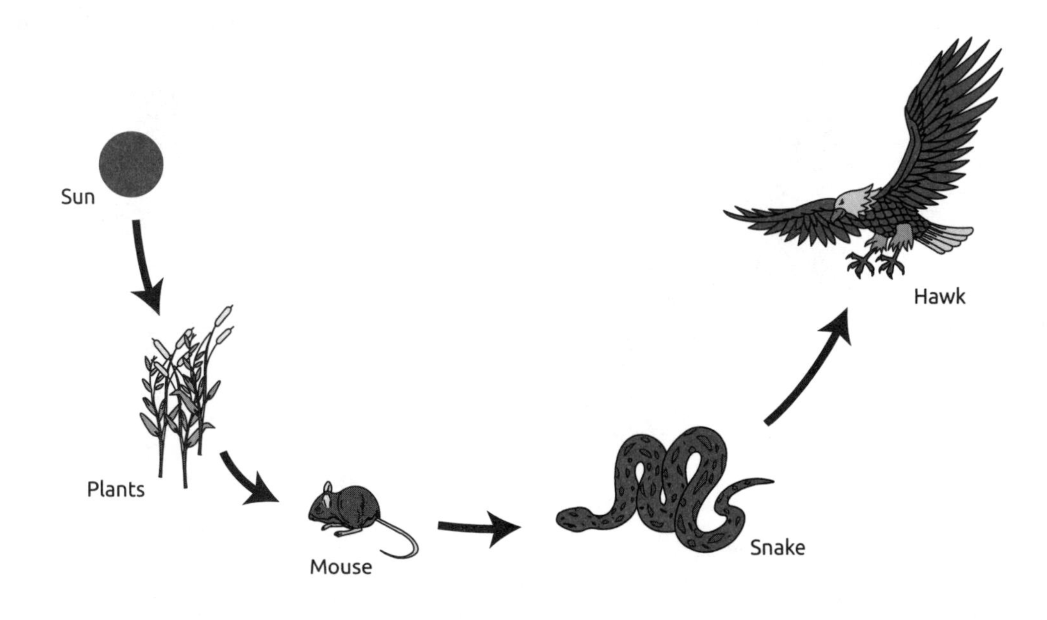

SYMBIOSIS

Key Terms

commensalism

mutualism

parasitism

symbiosis

J. R. R. Tolkien once wrote, "All have their worth and each contributes to the worth of others." He was referring to ecology. In nature and in life, each living organism is dependent on others. The changes in one population affect other populations. In this way, all things are connected.

Symbiotic Relationships

Organisms that live in the same environment have relationships with one another. Think about the relationships that may exist in a household. Some relationships are mutually beneficial, as when one person cooks a meal and another cleans up afterward. Some relationships benefit only one party, as when a parent makes a child's lunch each day. The child is benefitting, but the parent is not. Relationships may be negative, with one person benefitting while having a negative effect on another. An example could be a teenager who asks parents for money and clothes but never helps with chores.

These relationships are found in nature as well. **Symbiosis** is any relationship between two organisms. **Mutualism** is a relationship between two organisms in which both organisms benefit. Flowers and bees demonstrate mutualism. Bees gather nectar from flowers, which they use to make food. As the bees land on the flowers, they collect pollen on their bodies. The pollen is then transferred from one flower to another. This pollination allows the flower reproduce. Both the flower and the bee benefit from their relationship.

Commensalism is a relationship that benefits one organism, and the other is neither harmed nor helped. Barnacles may attach themselves to the skin of a whale. The barnacles are then transported to different areas as the whale swims about. In this way, the barnacles are able to obtain new sources of food. The whale does not receive any benefit but is not harmed either.

Some relationships help one organism and cause harm to another. These are parasitic relationships, or **parasitism**. Tapeworms are flatworms that attach themselves to the intestines of animals. The tapeworms eat the digested food within the animal, depriving the animal of nutrients. Fleas are another example of parasites. They attach themselves to the skin of dogs and are provided with shelter. They bite the dog's skin and drink the dog's blood to provide themselves with nutrients. The dog experiences pain and discomfort and loses small amounts of blood.

The image below shows a tapeworm living inside its host.

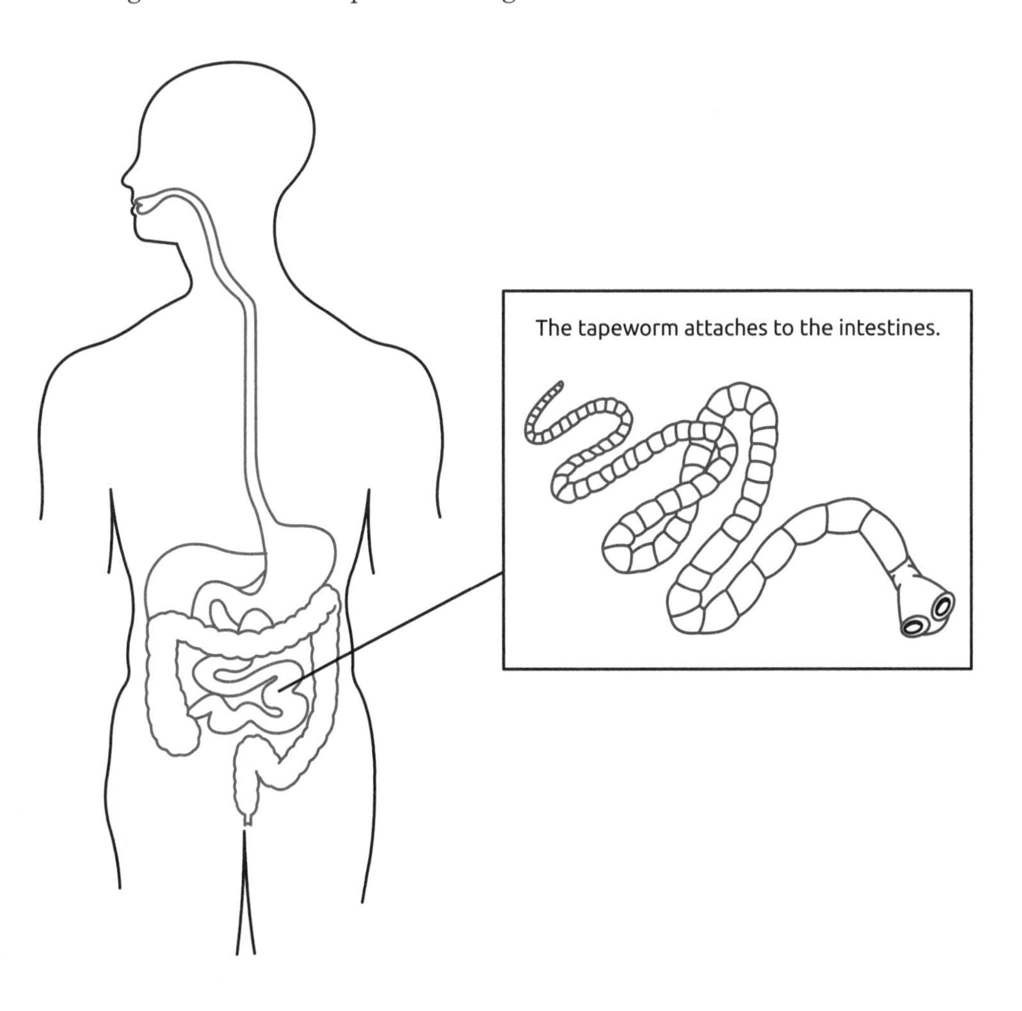
The tapeworm attaches to the intestines.

Predator-Prey Relationships

Predators are organisms that eat other organisms. Prey are the organisms that are eaten. Think about a bear eating a fish. The bear is the predator and the fish is the prey. These organisms have a relationship that is dependent on the other. If all the fish in a river were killed by disease, the bear would need to find a new type of prey to eat as it relies on fish for survival.

As a predator population changes, so does its prey population, and vice versa. Ladybugs eat aphids. If the number of ladybugs greatly decreased, what would happen to the number of aphids? The aphid population would increase, as they would be less likely to be eaten with fewer predators around. If the number of ladybugs increased, the number of aphids would decrease. If the number of aphids increased, the ladybug population would increase as well. Changes in a predator or prey population affect one another.

Real-World Connection

The Canadian lynx and snowshoe hare are a popular example of the predator-prey relationship. The lynx is the predator and the hare is the prey. Their populations were graphed over many decades. About two years after the population of the hare increased, the population of the lynx increased. The following year there was a sharp decrease in the population of the hares.

LESSON REVIEW

Complete the activities below to check your understanding of the lesson content. The Unit 2 Answer Key is on page 153.

Skills Practice

Answer the questions based on the content covered in the lesson.

1. Algae live on the backs of spider crabs. The algae get shelter. The algae provide the spider crab with camouflage. Which type of relationship is described?

   A. commensalism

   B. mutualism

   C. parasitism

   D. predator-prey

2. Tuna eat mackerel. If the tuna population decreased because of disease, what effect would this likely have on the mackerel population?

   ____

   ____

3. The caterpillar of the monarch butterfly eats milkweed plants. Milkweed contains a poisonous chemical that does not affect the monarch butterfly. The butterfly stores the chemical in its body, and if a bird or other predator eats the monarch, it will get sick. Predators learn to avoid the monarch butterfly. Identify and explain the type of relationship between monarch butterfly and milkweed plants.

   ____

   ____

   ____

4. In the following diagram of a food chain, which is an example of a predator-prey relationship?

   A. bobcat-algae

   B. algae-fish

   C. dragonfly-raccoon

   D. raccoon-fish

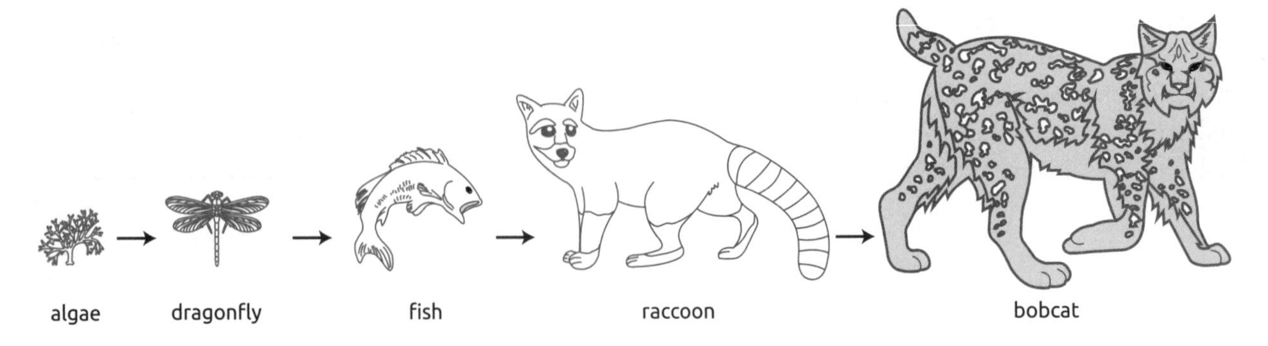

UNIT PRACTICE TEST

Answer the questions based on the content covered in this unit. The Unit 2 Answer Key is on page 153.

Answer the following questions.

1. Which of the following is correct about the cell membrane?

   A. It allows all substances to enter and leave the cell.

   B. It allows all substances to enter the cell, but only certain ones can leave.

   C. It allows only certain substances to enter the cell, but all substances can leave.

   D. It allows only certain substances to enter and leave the cell.

2. The following image signifies the function of which class of materials?

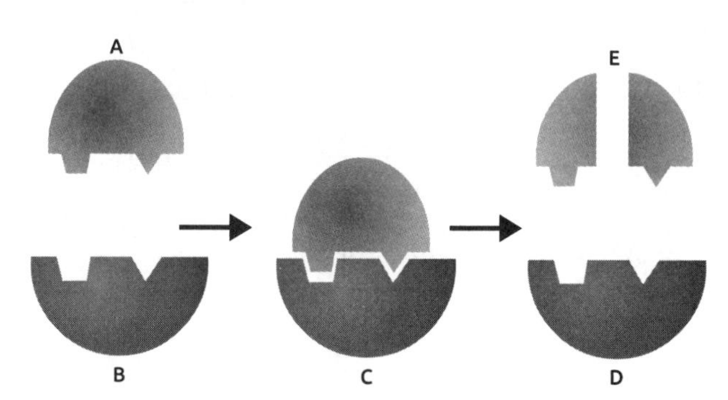

   A. enzymes

   B. minerals

   C. vitamins

   D. carbohydrates

3. Write the correct examples for each type of biological organization.

   examples: bone, white blood, heart, respiratory, red blood, muscle, pancreas, immune

cell		
tissue		
organ		
system		

Determine whether each statement describes mitosis or meiosis.

4. A single cell divides into four daughter cells.

   ____

5. The body uses it for growth or repair of damaged cells. ____________________

6. The new cells are identical to the parent cell.

   ____

7. Eggs and sperm are created through this form of division. ____________________

Select the correct option.

8. Groups of three bases on an RNA strand are called (chromosomes/codons).

9. An RNA strand is made by copying part of a DNA strand during (translation/transcription).

10. (Amino acids/Proteins) are long molecules consisting of many (amino acids/proteins) connected together.

Use the following information to answer questions 11 and 12:

Part of an RNA strand has this sequence:
UUCCACGUCGGCAUC

It produces this sequence of amino acids:
phe-his-val-gly-ile

11. What is the original DNA sequence?

    ____

12. What sequence of amino acids would be made by the DNA strand CAGAAGTAG?

    ____

Answer the following questions.

13. Draw a line to match the term to the example from the description. Not all examples will be used.

Term	**Example**
phenotype of offspring	YY
	yellow seed
genotype of offspring	Y
	YG
dominant gene	G
recessive gene	green seed

14. Two pea plants are bred to learn about plant height. They have offspring with the genotypes listed below.

    Tt

    tt

    TT

    The genotype of one parent is Tt. What is the genotype of the other parent? Explain how you know.

    ____

    ____

    ____

15. How does the epigenome prevent cells from performing incorrect functions?

    ____

    ____

    ____

16. Humans all have the same set of genes. How do variations among human phenotypes occur?

    ____

    ____

    ____

Read the following passage and then answer questions 17–19.

Walking sticks are long, thin insects that live on bushes. They have no wings and therefore spend most of their lives on one bush. One species of walking stick, *Timema cristinae,* lives in the hills of southern California. Scientist Patrik Nosil and his team studied two color variations of these insects: ones that are solid green, and ones with white stripes along the length of their bodies.

The solid green walking sticks live on bushes with thick green leaves, while the striped walking sticks live on bushes with needle-like leaves. Nosil and his team discovered that if the walking sticks were well matched to their bush, then they had a high population. If they were poorly matched, then their numbers were much lower.

17. How did the same species of insect come to have very different colorations?

    A. The insects came from different ancestors.

    B. The food the insects ate influenced their colors.

    C. Predation patterns of birds in the area were affected by the type of bush the insects lived on.

    D. Natural selection favored the adaptations of the insects that were well suited to the environment.

18. How could the researchers know if the two types of *Timema cristinae* became two different species?

    ____

    ____

    ____

19. What causes the variation of color patterns in *Timema cristinae*?

    A. different alleles

    B. a common ancestor

    C. different phenotypes

    D. an epigenetic signal

Answer the following questions.

20. Select ALL the options that are produced during photosynthesis.

    A. carbon dioxide

    B. sugar

    C. protein

    D. oxygen

21. All of the following are correct about trophic structures EXCEPT which one?

    A. Decomposers are part of trophic structures.

    B. An energy pyramid displays the flow of energy in trophic structures.

    C. Omnivores such as humans create imbalance in trophic structures.

    D. Trophic structures show the feeding relationship between organisms.

Read the following passage and then answer question 22 and 23.

Kudzu is a perennial vine of the legume family, native to Japan and southeast China. It was first brought to the United States in 1876. In the 1930s through the 1950s, the US government encouraged the planting of kudzu for soil erosion control. It was planted in abundance throughout the southern United States.

But kudzu, it turns out, grows rampantly, taking over native trees, road signs, and buildings. In 1953, kudzu was removed from the US Department of Agriculture's list of permissible cover plants due to its recognition as a pest species. In Florida, kudzu has currently been documented in 14 counties and is listed as a Category I invasive species. It is estimated that 2 million acres of forestland in the southern United States is covered with kudzu.

22. Which characteristic of kudzu makes it an invasive species?

    A. It is part of the legume family.

    B. It is a good plant for feeding livestock.

    C. The US government encouraged planting for soil erosion control.

    D. It takes over trees, road signs, and buildings.

23. Which is a condition in the United States that might have contributed to kudzu becoming invasive?

    A. a lack of livestock feed

    B. an overabundance of similar plants

    C. a lack of native forest animals that can eat kudzu

    D. an overabundance of highways

Read the following passage and then answer question 24 and 25.

Early on in its life, a sea anemone attaches itself to the shell of a hermit crab. Both creatures tend to grow at approximately the same rate. When the hermit crab outgrows its shell, the sea anemone travels right along, attaching itself to the crab's new shell. Both creatures benefit from each other's presence in a number of different ways. For example, the sea anemone protects the crab from predators. When sensing danger, the hermit crab communicates with the sea anemone, which then extends thread-like tentacles that will sting the predator. In return, the hermit crab allows the sea anemone to eat its leftovers. Because the hermit crab is mobile and moves from place to place in search of food, the sea anemone is also guaranteed a steady supply of food.

24. Which best describes the relationship between the hermit crab and sea anemone?

    A. commensalism

    B. mutualism

    C. parasitism

    D. predator-prey

25. Which would most likely happen if the hermit crab population decreased?

    A. The sea anemone population would increase.

    B. The sea anemones would die out.

    C. The sea anemone population would decrease.

    D. The sea anemones would become more mobile.

Answer the following question.

26. Which best describes an example of how the endocrine system works with other organ systems?

    A. Growth hormones stimulate cell division in the bones.

    B. Chemical messengers produced by the thyroid gland send signals to other glands.

    C. Red blood cells are transported by arteries to the digestive organs.

    D. Gases taken in by the respiratory system travel to the cells of the body.

UNIT 3 Physical Science

Energy is all around us! We can hear it when our friends talk to us, we can see it when the sun shines down on us, and we can feel it when the wind blows. We also use energy all the time—for example, when we walk, play video games, cook food, or drive a car.

Unit 3 Lesson 1 TYPES OF ENERGY AND TRANSFORMATIONS

Key Terms

energy

gravitational potential energy

kinetic energy

Law of Conservation of Energy

mechanical energy

potential energy

work

Energy is the ability to do work. The scientific definition of **work** is to use force to move something over a distance. Energy comes in many different forms, but all energy can be classified as kinetic or potential.

Kinetic Energy

Kinetic energy is the energy of moving objects. Sound is made of moving sound waves, so it is a type of kinetic energy. Electricity is another form of kinetic energy because it is made up of moving electrical charges.

The faster an object moves, the more kinetic energy it has. The greater the mass of a moving object, the more kinetic energy it has. You probably already know that a car moving fast will have more energy than a slow car, and a big truck will have more energy than a small car. This is because kinetic energy is dependent upon both mass and speed.

TYPES OF ENERGY AND TRANSFORMATIONS

Potential Energy

Potential energy is stored energy. Chemical energy is an example of potential energy. Chemical energy is the energy that holds elements together in a chemical compound. When a compound's chemical bonds are broken and new bonds are formed, energy is released. Gasoline and food both have chemical energy. Nuclear energy is a form of chemical energy. It is the energy that holds the nucleus of an atom together. When the nucleus is split apart and new nuclei are formed, a large amount of energy is released.

Potential energy can also be found in a stretched rubber band, a rock on the edge of a cliff, and an arrow pulled back in a bow. As with kinetic energy, the more mass an object has, the more potential energy it has.

Gravitational potential energy is a type of potential energy. This is energy that is stored in an object and is related to the height of that object. The energy is stored in the object because of the force of gravity exerted by Earth. The higher something is, the greater the gravitational potential energy. Imagine a boulder on a high cliff. Because of the force of gravity pulling it to the ground, it has the potential to do a lot of work if it falls from the cliff. So, the higher an object and the more mass it has, the greater its gravitational potential energy.

Mechanical Energy

As stated earlier, energy is required to do work. When work is done to an object, that object gains energy. This energy is called **mechanical energy**. For instance, when you throw a baseball, you give energy to the ball to allow it to move across the field. Another example is when a rolling bowling ball hits bowling pins. The ball transfers some of its energy to the pins, which allows them to move. Mechanical energy is determined by the total amount of kinetic and potential energy in an object.

Energy Can Change from One Form to Another

An important concept in science is called the **Law of Conservation of Energy**. This law states that energy cannot be created or destroyed. Another way of saying this is that energy is conserved. Although energy is conserved, it can change from one type to another. In other words, the total amount of energy stays the same, but it can be in different forms. When energy changes from one form to another, we say that it is "transformed" or "converted" to a new form of energy.

Energy Transformations

If you have ever been on a roller coaster, you have experienced energy conversions. When a roller coaster car is pulled up to the top of a hill, mechanical energy is used. When the car is stopped at the top of that hill, the mechanical energy has been transformed into potential energy. The higher the car is, the more potential energy it has. Because the car is not moving, it has no kinetic energy. As the car moves down the hill, however, it rolls faster, gaining kinetic energy as it speeds up. Because it is moving down the hill, its height is getting smaller, so the potential energy is decreasing. At the lowest point in the hill, it has little potential energy, but the car is moving at the fastest rate, so there is a lot of kinetic energy. As the car moves up another hill, it gains potential energy and loses kinetic energy, and the process repeats itself.

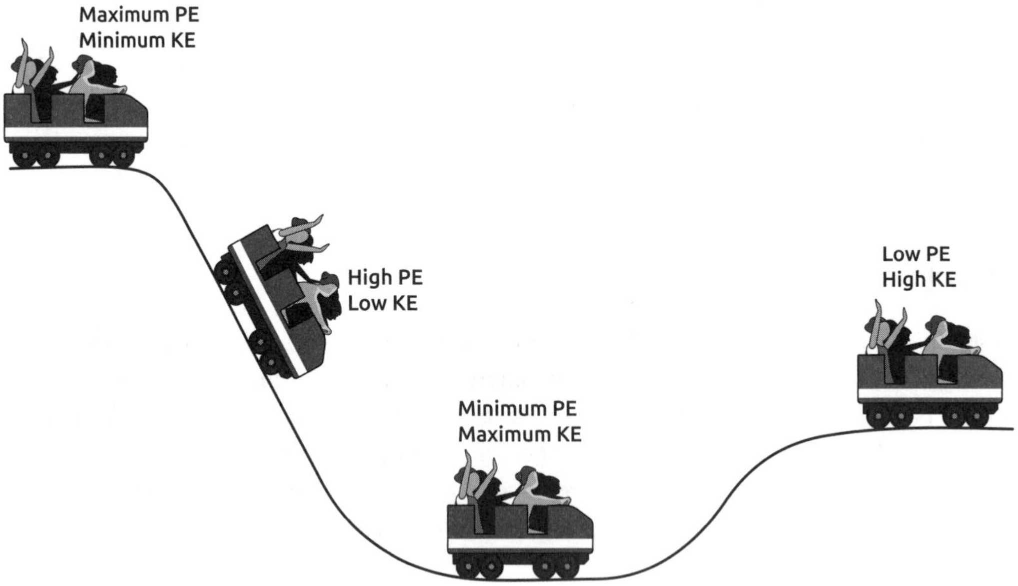

Maximum PE
Minimum KE

High PE
Low KE

Minimum PE
Maximum KE

Low PE
High KE

Skills Tip

To remember what potential energy is, think of what "potential" means. Something with potential has the ability to do something at a future time. An object with potential energy can do work at some point, but right now that energy is being stored.

A battery contains chemical energy. When you use a battery-powered device, the chemical energy is converted to electrical energy to run the device. In a car engine, fuel is burned. The chemical energy stored in the chemical bonds in the fuel is transformed into heat. The heat energy is then converted to mechanical energy to move the car. Burning fuels also produces light energy, such as when the wax in a candle burns. Another type of energy transformation occurs when you turn on a light. As the electricity moves through the light bulb, it is transformed into light energy.

LESSON REVIEW

Complete the activities below to check your understanding of the lesson content. The Unit 3 Answer Key is on page 154.

Skills Practice

Answer the questions based on the content covered in the lesson.

1. Describe the relationship between potential, kinetic, and mechanical energy.

____

____

Fill in the blank.

2. A stretched-out spring has ____________________ energy.

Select the best option.

3. The energy of movement is (kinetic, potential) energy.

4. Stored energy is (kinetic, potential) energy.

5. A bird flying through the air has (potential, kinetic) energy.

6. When fuel is burned, the chemical energy in the fuel is converted to which kinds of energy?

   A. potential and electrical energy

   B. electrical and light energy

   C. heat and potential energy

   D. heat and light energy

7. Which type of energy is released when the bonds in sugar are broken to form carbon dioxide and water?

   A. chemical

   B. electrical

   C. kinetic

   D. mechanical

8. Which energy transformation occurs when you turn on a fan and it causes the air to blow?

   A. potential to electrical

   B. chemical to potential

   C. mechanical to chemical

   D. electrical to mechanical

SOURCES OF ENERGY

Key Terms

fission

fossil fuel

generator

greenhouse gas

hydroelectric power

nonrenewable resource

nuclear energy

renewable resource

turbine

Real-World Connection

When fossil fuels are burned, they release gases, including carbon dioxide. Carbon dioxide is a natural part of our atmosphere, but as more fossil fuels are used, there is more carbon dioxide in the atmosphere. Carbon dioxide is a greenhouse gas. **Greenhouse gases** hold heat in Earth's atmosphere, which results in global warming. Global warming creates a variety of changes in the world's climate, which is why the term "climate change" is more accurate than "global warming."

In order to power our appliances, computers, and lights, we need electricity. The electricity in our homes comes from different sources, depending on where we live. Most homes rely on fossil fuels, but other forms of energy are becoming more common as well.

Renewable and Nonrenewable Energy Sources

Renewable resources will never run out. Renewable resources include solar energy and the energy from moving air (wind) and water. **Nonrenewable resources** can be used up. Fossil fuels are an example of a nonrenewable energy source.

Fossil Fuels

Fossil fuels were formed over millions of years from the remains of dead plants and animals (giving them the name "fossil" fuels). Because they form so slowly, they can't be replaced once they are used, so fossil fuels are considered nonrenewable. Fossil fuels contain chemical energy, which can be converted into other types of energy. In fact, this is what happens in a power plant.

In a power plant, the chemical energy in the bonds of natural gas, oil, or coal is converted to heat energy. The heat is used to produce steam. The steam is then directed to a turbine. A **turbine** is a machine that spins as a result of some gas or fluid flowing past it. As the steam passes over the turbine, it rotates at a high speed. The turbine now has mechanical energy. The mechanical energy of the turbine is transformed into electrical energy in a **generator**.

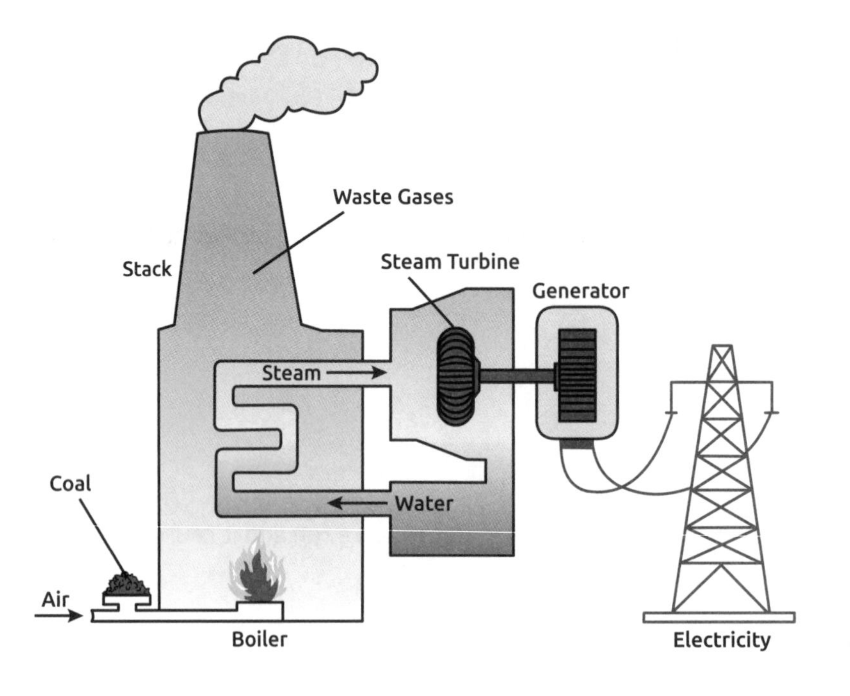

Nuclear Energy

Nuclear energy is another power source used to generate electricity. Nuclear energy is a form of chemical energy, like fossil fuels. Nuclear energy is the energy that holds the nucleus of an atom together. If the nucleus of an atom is split and two or more new nuclei are formed, a large amount of energy is released. The process of splitting an atom is called **fission**. In nuclear power plants, atoms of the element uranium are split, releasing energy. That energy is used to heat water, which creates steam. The steam is then used to turn a generator, such as those in coal-burning plants.

Renewable Energy Sources

Renewable energy sources, such as the sun, wind, and water, can never run out. Because wind is moving air, it contains mechanical energy, which can then be changed into a different form of energy. The same is true for moving water. As in the coal-burning power plant shown on the previous page the energy from wind or water is used to turn a turbine. The turbine turns a generator, transforming mechanical energy into electric energy. For example, in a **hydroelectric power** plant, the flow of water turns the turbine. These plants are usually set up where they can take advantage of the greatest fall of water, as in a deep valley.

Solar power works differently. Solar panels, like the ones you see on rooftops and calculators, convert the sun's energy into electrical energy. Sunlight is made up of tiny energy particles called photons. As the photons hit the atoms of the solar panels, they give extra energy to some of the atoms' electrons. This causes them to break away from the atoms. The structure of the panel forces the loose electrons to flow in the same direction. Flowing electrons form an electric current, providing electricity to the device or home.

Advantages and Disadvantages of Different Energy Sources

As you have seen, a number of energy sources can be used to create electricity. Countries around the world are working toward the use of more renewable forms of energy. The following table shows the advantages and disadvantages of the different forms of energy used today.

Energy Source	Advantages	Disadvantages
Nuclear energy	• Inexpensive fuel • No smoke particles to pollute the air • Lots of energy produced from a small amount of fuel • Not dependent on weather	• Produces radioactive waste that is dangerous to humans and the environment • Plants are expensive to build. • Nuclear accidents can release radiation that can make living things sick or even die.
Fossil fuels	• Relatively easy to access • Easy to store and transport • Plants are relatively inexpensive to build.	• Will eventually run out • Contribute to climate change • Release smoke and chemicals, which pollute the air
Renewable energy sources: sun, wind, water	• Cannot be used up • Cause very little pollution • Do not contribute to climate change	• New technology can be expensive to implement. • Not as efficient as fossil fuels, so more energy facilities are needed • Unpredictable because of dependence on weather

There are advantages and disadvantages to each form of energy, but it is important to take a long-term view of energy use. We have to decide which sources are best for our planet and best for our future.

LESSON REVIEW

Complete the activities below to check your understanding of the lesson content. The Unit 3 Answer Key is on page 154.

Skills Practice

Answer the questions based on the content covered in the lesson.

1. Describe two advantages of using fossil fuels.

___

___

2. Describe two advantages of using renewable resources.

___

___

Select the best option.

3. Which best describes the difference between renewable and nonrenewable resources?

   A. Renewable resources never run out.

   B. Renewable resources come from nature.

   C. Renewable resources contain less energy.

   D. Renewable resources are always available for use.

4. Which of the following can release the greatest amount of energy using the smallest amount of the resource?

   A. gas

   B. oil

   C. uranium

   D. wind

Fill in the blanks.

5. One problem with ______________ power is the storage of radioactive waste.

6. ______________ from the sun are responsible for generating electrical energy in solar panels.

7. One problem with renewable power sources is that they are somewhat dependent on ______________.

8. On a wind farm, moving air spins a ______________ to generate electricity.

FLOW OF ENERGY

Key Terms

conduction

convection

endothermic reaction

exothermic reaction

heat

infrared radiation

radiation

temperature

thermal energy

Real-World Connection

If you have ever tried to boil a large pot of water, you have seen the difference between heat and temperature. A large pot of water takes a lot longer to boil than a small cup of water does. This is because the greater mass of water in the pot requires more heat energy to get hot.

Energy comes in different forms, including sound, light, and electricity. In this lesson, we will focus on heat energy. We are all familiar with heat energy, but how can we measure heat, and how does heat energy behave?

Heat and Temperature

Heat and temperature are related, but they are not the same thing. **Heat** is a type of energy, so it can do work. Heat is often referred to as **thermal energy**. **Temperature** is a measure of how hot something is, or how much kinetic energy its particles have. The faster the particles of a substance move, the more kinetic energy they have. Substances with more kinetic energy have a higher temperature. The amount of heat energy in a substance depends on the kinetic energy of its particles, as well the number and type of particles. This is why two different substances with the same mass can be at the same temperature but have different amounts of heat energy.

How Does Heat Move?

Generally speaking, things tend to flow downward, until everything is as close to the same level as possible. For instance, if you pile sand up into a hill, some will roll back down; the same is true for heat energy. Areas of high temperature give off energy to areas with lower temperatures. For example, if you heat one part of a pot, the heat will move toward the cooler parts of the pot to even out the temperature. In the warmer part of the pot, the particles with more heat energy vibrate at a rapid speed. As they vibrate, they bump into particles nearby and make them vibrate more. This passes the thermal energy through the pot from the warm area to the cooler area. This process is called **conduction**. Conduction involves the transfer of heat through direct contact. **Convection** describes the way heat moves within a liquid or gas. When air is warmed, its particles move faster (they have more kinetic energy) and spread out. This makes the air lighter, or less dense; less dense gases will rise above denser, cooler gases. As the warm air rises, cooler air moves in to take its place; this movement of air creates a cycle that works to warm up a larger area. The process works the same way for liquids.

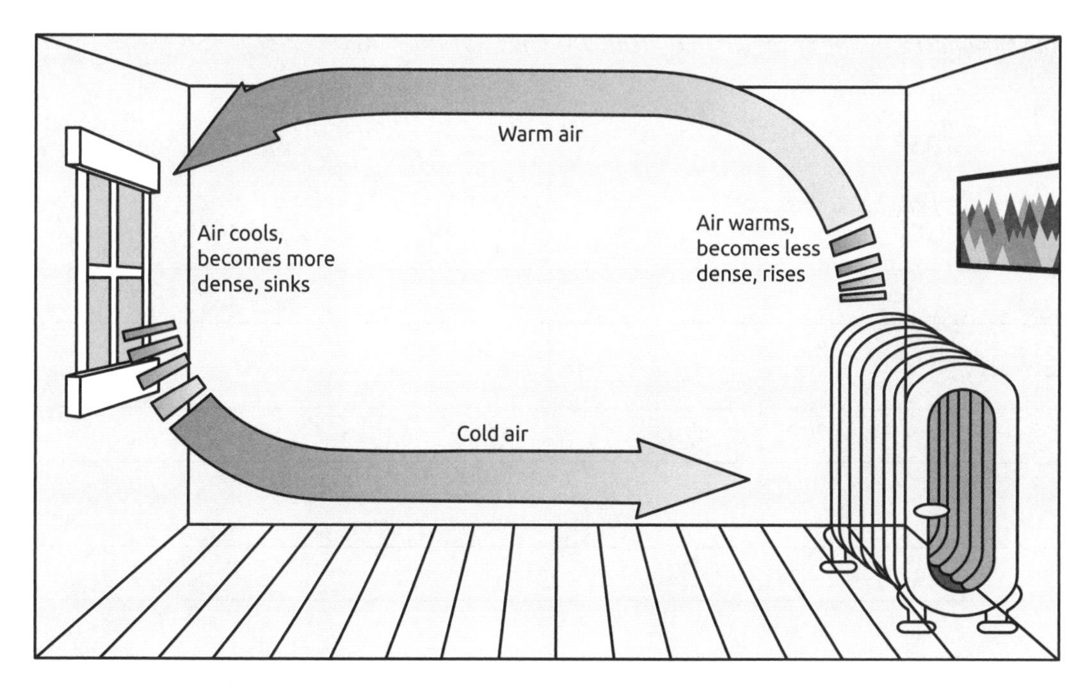

Convection in a room

Radiation is a third form of heat energy movement. Unlike conduction and convection, particles are not involved in this type of heat transfer. This is why we can feel the heat of the sun. Thermal radiation is also called **infrared radiation**. Hotter objects release more infrared radiation. Night vision goggles use infrared detection to "see" the bodies of humans in the dark because they are generally warmer than their surroundings.

Heat and Chemical Reactions

You may remember that chemical energy is the energy held in the bonds of chemical compounds. Breaking these bonds requires the absorption of energy, and forming new bonds releases energy. If more energy is absorbed than released, it is an **endothermic reaction**. Endothermic reactions absorb heat from their surroundings. This is useful for things like chemical ice packs, which are activated by squeezing the bag to release and mix the chemicals. The reaction that occurs to make the ice packs cold is endothermic.

If, instead, more energy is released by the reaction than is absorbed, it is an **exothermic reaction**. Exothermic reactions release heat to their surroundings. Burning and rusting are examples of exothermic reactions. Ready-to-eat meals are made with packaging that contains chemicals that undergo an exothermic reaction when activated. The food inside heats up as the reaction releases heat into the packaging.

LESSON REVIEW

Complete the activities below to check your understanding of the lesson content. The Unit 3 Answer Key is on page 154.

Vocabulary

Write definitions in your own words for each of the key terms.

1. conduction ____

____

2. convection ____

____

3. thermal energy ____

____

4. temperature ____

____

Skills Practice

Answer the questions based on the content covered in the lesson.

5. Which option correctly compares a large container of water at 30 degrees Celsius and a small container of water at 30 degrees Celsius?

   A. The large container has more heat energy.

   B. The small container has more heat energy.

   C. The large container has a higher temperature.

   D. The small container has a higher temperature.

6. Which type of heat transfer can occur in space?

   A. conduction

   B. convection

   C. radiation

   D. all of the above

Select the best option.

7. An (endothermic, exothermic) reaction requires energy.

8. A student conducts a chemical reaction. As it is occurring, the student notices that the test tube feels warm. The reaction was (exothermic, endothermic).

9. A reaction that uses less energy to break old bonds than the energy it releases in making new bonds is (endothermic, exothermic).

10. The warmer an object, the more (chemical, kinetic) energy it has.

The Nature of Radiation

Rattlesnakes seem to have an unfair advantage when hunting for prey. They are able to sense the heat from living creatures using infrared detection. Rattlesnakes are essentially wearing night vision goggles for all their hunts. This allows them to spot their prey. Energy from body heat is not visible to people without special devices. Most energy is not visible to humans.

Energy is emitted in many different forms. **Radiation** is the release of energy as an electromagnetic wave. All forms of radiation travel at the speed of light. The various forms of radiation form the **electromagnetic spectrum**. This is the entire range of radiation. It ranges from radio waves to gamma rays. Each type has a different frequency and wavelength. **Frequency** refers to how often the particles vibrate as a wave passes through a medium. **Wavelength** is the distance between two successive points in a wave. Human eyes cannot see most waves on the electromagnetic spectrum.

The following diagram shows the electromagnetic spectrum.

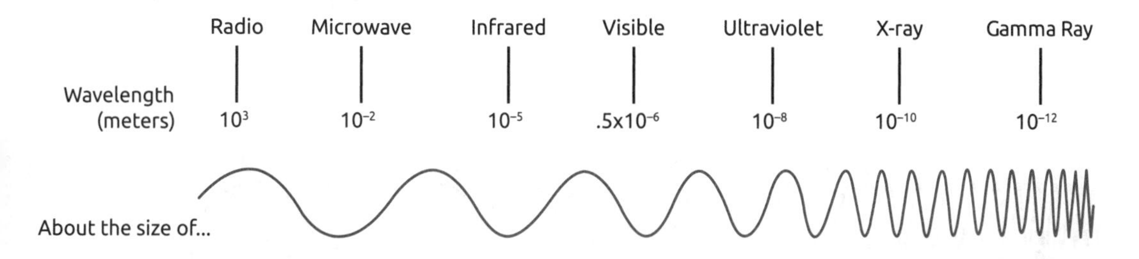

Key Terms

electromagnetic spectrum

frequency

radiation

wavelength

The Electromagnetic Spectrum

Each type of radiation on the spectrum has different characteristics and uses. Radio waves have the longest wavelengths and a very low frequency. Radio waves cannot be seen by the human eye. They are able to pass through the body without causing any damage. Radio waves are used to transmit television and radio signals. Cell phones also use radio waves.

Microwaves have shorter wavelengths than radio waves and a higher frequency. They cannot be seen. A microwave oven uses microwaves to heat the temperature of food and drinks.

Infrared light waves can be felt as heat, but it cannot by seen without night vision goggles. Infrared light is used in some restaurants to keep food warm. Some remote control devices also use infrared energy.

Visible light is the only light waves that can be seen by the human eye. The colors that we see are made up of the visible light spectrum and differ in their wavelength. Red has the longest wavelength, followed by orange, yellow, green, blue, indigo, and finally violet, with the shortest wavelength. Visible light is used in photography and human vision.

Ultraviolet is the next type of electromagnetic radiation. The sun emits ultraviolet (UV) rays to Earth. They cannot be seen or felt, but they can cause our skin to burn. It is important to wear sunblock to protect your skin from UV rays. These rays are used in fluorescent lighting and for security markings.

X-rays have a very small wavelength and high frequency. They are able to penetrate our skin but cannot pass through bones, teeth, or metal. X-rays are used frequently in the medical and security fields.

Gamma rays have the shortest wavelengths. They have a very high frequency. They cannot be seen by the human eye. Gamma rays are used to sterilize equipment and to target and kill cancerous cells.

Real-World Connection

Ultraviolet light is used to help detect counterfeit money and identification cards. The cards and bills are printed with special inks. When a bill is placed under a UV light, the design can be seen. Bills or cards that do not have this design can be identified as counterfeit.

Dangers of Radiation

Electromagnetic radiation can be dangerous. The higher the frequency of the wave, the more damage it can cause. Ultraviolet rays can cause sunburn to your skin. X-rays can damage cells and cause mutations, which may lead to cancer. Gamma rays can be used to kill cancer cells, but they can also kill and damage other types of cells. It is important to limit exposure to radiation.

Unit 3 Lesson 4 LESSON REVIEW

Complete the activities below to check your understanding of the lesson content. The Unit 3 Answer Key is on page 154.

Vocabulary

Write definitions in your own words for each of the key terms.

1. electromagnetic spectrum ______________________________

____

2. frequency ______________________________

____

3. radiation ______________________________

____

4. wavelength ______________________________

____

Skills Practice

Answer the questions based on the content covered in the lesson.

5. This type of radiation has the longest wavelengths and a low frequency. It is used to transmit TV signals. Which type of radiation is this?

   A. gamma

   B. radio

   C. visible

   D. x-ray

6. Which characteristic could describe x-rays?

   A. high frequency

   B. long wavelength

   C. used in fluorescent lighting

   D. used to heat food and drinks

7. In the following list, choose the option that fits in the electromagnetic spectrum in order of low to high frequency:

   radio wave, visible light, (infrared light, microwave, ultraviolet), gamma ray

MOTION

Key Terms

acceleration

collision

momentum

speed

vector quantity

velocity

What Is Motion?

Stand in front of a mirror and try to remain as still as possible. Try not to move any part of your body. It is harder than it sounds. This is because our bodies are almost always in some sort of motion, no matter how slight. Motion, or movement, can be described by many terms. Think about a speed limit sign you see when driving. This sign communicates the highest speed a person should drive in an area. **Speed** is the rate at which an object moves. It is computed by dividing distance by time. Often, speed is expressed in units of m/s, or meters per second.

Imagine a road sign that sets the speed limit at 50 miles per hour (mph). This means that if you drive at a speed of 50 mph for one hour, you will travel 50 miles.

The Movement of a Runner

It takes a student 15 seconds to get up to her top running speed; once she is at this speed, she can run 100 meters in 15 seconds. To calculate this runner's speed, divide the distance traveled by the time. The distance was 100 meters, and she ran it in 15 seconds:

$$\frac{100 \text{ m}}{15 \text{ s}} = 6.67 \text{ m/s}$$

The speed the student ran was 6.67 m/s.

Velocity is a term often used to describe objects in motion. **Velocity** is the speed of an object in a given direction or the rate at which an object changes position. Velocity is a **vector quantity**, which means that the magnitude and direction must be included. In the previous example, the velocity could not be determined since no direction was specified. Say the student runs due west 100 meters in 15 seconds. The velocity of the runner is 6.67 m/s, due west.

Now you can find the acceleration of the runner. **Acceleration** is the rate at which an object changes velocity. It is calculated by dividing the change (Δ) in velocity by time:

$$a = \frac{\Delta\text{velocity}}{\text{time}} = \frac{v_f - v_i}{t}$$

If the velocity of an object does not change, the object has no acceleration. In the example of the student running, the runner initially was standing still, or moving at 0 m/s.
It took her 15 seconds to get up to her final constant velocity of 6.67 m/s. To find the acceleration, subtract the initial velocity (0 m/s) from the final velocity (6.67 m/s), and divide this by the length of time it took her to get up to her final velocity:

$$a = \frac{6.67 \text{ m/s} - 0 \text{ m/s}}{15 \text{ s}} = \frac{6.67 \text{ m/s}}{15 \text{ s}} = 0.44 \text{ m/s}^2$$

The runner increased velocity at 0.44 m/s^2. Acceleration is a vector quantity, so it is described by magnitude and direction. Acceleration should include a direction indicating whether the velocity increases or decreases.

Momentum and Collisions

If a runner is attempting to jump as far as possible, is it better for him to have a running start or to start from a standstill? Usually, if an object is in motion going into a jump, then it will have greater momentum. An object's **momentum** is defined as the motion of the object. It is calculated by multiplying the mass and velocity of an object:

momentum $(p) = m \times v$

The direction of the object should be included in the momentum because it is a vector measurement. Two balls are rolling with equal velocity of 2 m/s westward. Ball A has a mass of 0.15 kg. Ball B has a mass of 7 kg. Which ball has a greater momentum?

Ball A: p = 0.15 kg × 2 m/s = 0.3 kg · m/s

Ball B: p = 7 kg × 2 m/s 14 kg · m/s

Both balls would be moving in a westward direction. Ball B has a greater momentum because the mass is larger, and mass has a proportional relationship with momentum.

A **collision** occurs when two or more objects collide, or run into each other. Both objects were in motion and had momentum prior to the collision, so the total momentum of the objects is conserved. If one person is standing still and another person trips and bumps into him, what happens to the first person? Does he remain standing still or does he move forward, while the person who tripped remains where she bumped into him? The person who was collided with will move forward due to the conservation of momentum and energy. The person who tripped transferred her forward momentum onto the person who was standing still. Collisions and transfers of energy can occur in many different ways.

Real-World Connection

Momentum is a term frequently used in the sports world. A team could be described as having great momentum going into playoffs. Momentum means the motion of an object, and the team is not physically in motion. This metaphor actually means that the team has strength going into the playoffs and is going to be difficult to beat or stop, as an object with great momentum would be.

LESSON REVIEW

Complete the activities below to check your understanding of the lesson content. The Unit 3 Answer Key is on page 154.

Vocabulary

Write definitions in your own words for each of the key terms.

1. acceleration ____________________
2. collision ____________________
3. momentum ____________________
4. speed ____________________
5. vector quantity ____________________
6. velocity ____________________

Skills Practice

Answer the questions based on the content covered in the lesson.

7. A car travels 195 miles in 3 hours. What is the average speed of the car?

   A. 60 mph

   B. 65 mph

   C. 192 mph

   D. 198 mph

8. A car travels 195 miles in 3 hours. What information needs to be added in order to determine the velocity of the car?

   A. acceleration

   B. direction

   C. mass

   D. time

FORCES

One day in 1666, young Isaac Newton was sitting beneath an apple tree. Suddenly, an apple fell on his head. According to the legend, with the thump of the apple came the realization of the theory of gravity. This story may have been embellished over the years by both Sir Isaac Newton and generations to follow. However, this simple event demonstrates gravitational force.

Key Terms

force

gravity

inertia

Newton's First Law of Motion

Newton's Second Law of Motion

Newton's Third Law of Motion

Forces

A **force** is a push or a pull on an object that results from that object's interaction with another object. A force can cause a change in motion. For example, if a box is sitting on the ground, and then you push the box across the room, you are applying a force to the box.

Opposing forces can be balanced or unbalanced. When they are balanced, the object they are acting upon does not move. If you are pushing on one side of the box and your friend is pushing on the opposite side with the same amount of force, the box will stay in place.

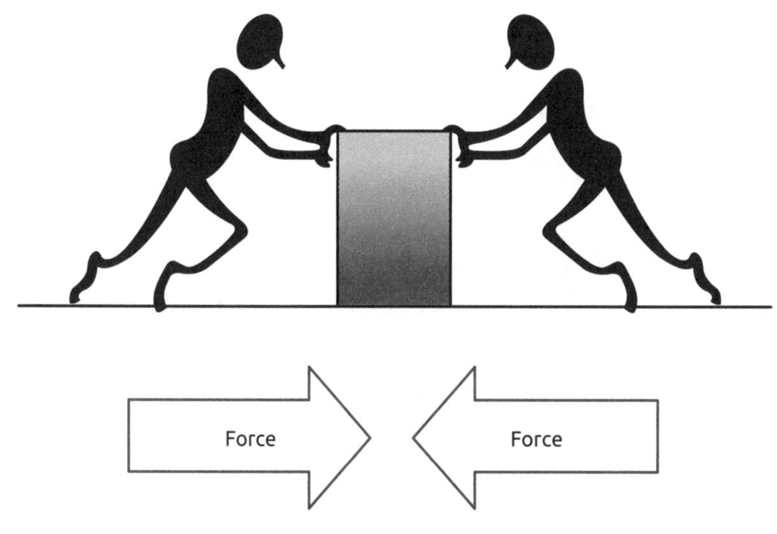

Balanced forces

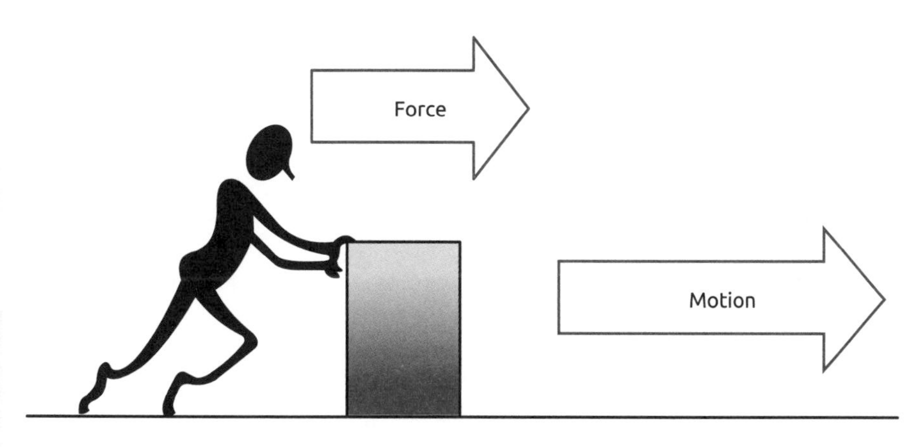

Unbalanced forces

FORCES

Most people are familiar with the force of **gravity**. On Earth's surface, all objects are pulled downward toward the center of the planet. When the apple fell on Newton's head, it was being pulled toward Earth's center. The force of gravity is determined by the mass of the two objects—in this case, Earth and the apple—and the distance between them. Weight is the force of a planetary body's gravity on a surface object. A smaller planetary body has weaker gravity. A person weighing 150 pounds on Earth would weigh 24.9 pounds on its less massive moon.

Real-World Connection

When a vehicle comes to a sudden stop, a passenger will continue to move due to inertia. This is why it is important to wear a seat belt. Without the restraint of a seat belt, the passenger will keep moving forward, possibly being ejected from the vehicle. The passenger may also suffer from whiplash.

Newton's First Law of Motion

The box sitting on the floor will stay there until you come along and push it. It will not move on its own. This is known as **inertia**, described by **Newton's First Law of Motion**. Inertia is the tendency of an object to stay at rest or stay in motion unless acted upon by an outside (unbalanced) force. If you roll a ball, it will eventually come to a stop because outside forces are acting on it—in this case, friction. If you roll that same ball on a frictionless surface, it will never stop.

Newton's Second Law of Motion

Newton's Second Law of Motion is demonstrated by the following formula:

F (force) = m (mass) × a (acceleration)

This means that the greater the mass of an object, the greater the force needed to move (accelerate) that object. It will take more force for you to push a heavier, more massive box across the floor than to push a lighter, less massive one.

Let's look at an example of how this formula can be used to find force.

Louisa has a box of books with a mass of 50 kg. How much force does she need to apply to the box to push it across the floor with an acceleration of 0.2 m/s^2? Assume there is no friction between the box and the floor.

We know the following:

$F = ?$

$m = 50$ kg

$a = 0.02$ m/s/s

Plug the numbers into the equation:

$F = (50 \text{ kg})\ (0.2 \text{ m/s/s})$

$F = 10 \text{ kg} \cdot \text{m/s/s}$, or 10 N (Newtons)

Newton's Third Law of Motion

Newton's Third Law of Motion states, "For every action, there is an equal and opposite reaction." In other words, for every force applied to an object, an equal force acts in the opposite direction. Imagine that you and a friend, who weighs the same as you, are both standing on skateboards. If you give your friend a push, *both* of you will move with the same amount acceleration, but in opposite directions.

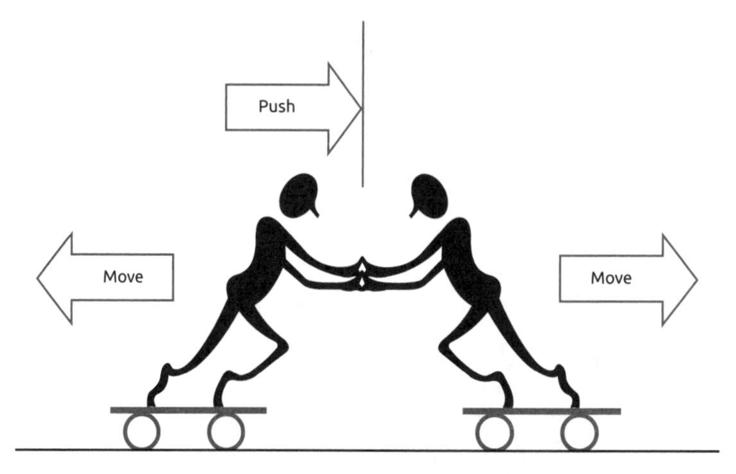

When you push your friend, you will both move with the same acceleration in opposite directions.

Unit 3 Lesson 6 LESSON REVIEW

Complete the activities below to check your understanding of the lesson content. *The Unit 3 Answer Key is on page 154.*

Skills Practice

Answer the questions based on the content covered in the lesson.

1. Joe is moving a 100-kg couch from one side of the room to the other. How much force does he need to apply to the couch to move it at an acceleration of 0.3 m/s^2? Assume there is no friction between the couch and the floor.

   A. 0.3 N
   B. 30 N
   C. 97 N
   D. 100 N

2. Darlene pushes on a box with a force of 2.5 N. Max pushes on the same box, at the same time, with a force of 1.2 N. Which statement best describes the forces acting on the box?

   A. The forces are balanced.
   B. The forces are unbalanced.
   C. The forces will cancel each other out.
   D. The forces will move the box toward Max.

3. The mass of Earth is 5.98×10^{24} kg. The mass of Mercury is 0.33×10^{24} kg. How would the weight of a person with a 150-kg mass compare on the two planets? Explain your reasoning.

   ____

   ____

4. Explain why a book resting on a table does not fall to the floor.

   ____

   ____

LESSON REVIEW

5. Andy and Javier are playing tug of war. The forces with which they are each pulling the rope are indicated in the diagram.

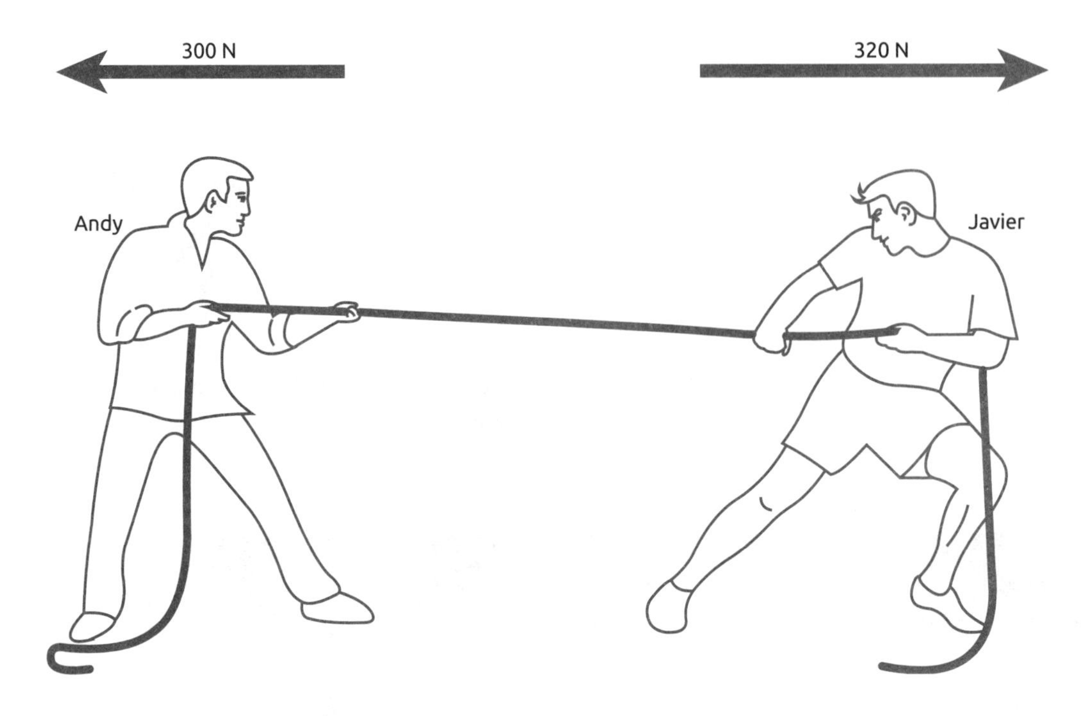

Describe how the rope will move in this tug of war and explain why this occurs.

____

____

6. A 100-kg car and a 500-kg truck are being crash-tested in a lab. Both vehicles are moving toward a wall with the same speed, and both take half a second to stop after hitting the wall. Which vehicle hit the wall with the most force? Explain why.

____

____

Fill in the blanks.

7. For every action, there is an equal and opposite ______________________.

8. A force can cause a change in ______________________.

Select the best option.

9. A more massive object has a (greater/lesser) inertia than a less massive object.

10. A planet with a large mass has a (greater/lesser) gravitational force than a planet with a small mass.

WORK

You've just completed preparations for a move: packing your belongings, lifting heavy boxes from your home, carrying them down the driveway, and then loading them up into a truck. Your heart is racing, and you wipe the sweat from your brow. As you reach for a cool drink of water, you think to yourself, *That was a lot of work.* Work did occur during this scenario, but in the physics world, what exactly is work?

Key Terms

complex machine
energy
inclined plane
joule
lever
machine
power
pulley
screw
watt
wedge
wheel and axle
work

Work, Energy, and Power

When you lift a heavy box, you are doing **work** upon that object, because a force (you pulling up on the box) acting on the box causes displacement of that box. For a force to qualify as having done work on an object, the object must be displaced; in other words, you can push against a wall for hours, but if the wall does not move, then no work has occurred. Work is measured in a unit called a **joule**.

In order to do work, you must have **energy**, the capacity for doing work. Without energy, you cannot apply a force, and no work can occur. The amount of energy expended to do work must be equal to or greater than the amount of work that is to be done. For example, to do 50 joules of work, you must expend 50 joules of energy. In reality, you must expend a little more because some energy will be lost as heat.

Power is the rate at which work is done, or work per unit of time. It can also be said that work is the rate of using energy. If one joule of work is done on an object in one second, then the amount of power produced is one **watt**. We often see power measured on electric meters as kilowatts (1,000 watts). Another common unit of power, most often seen in automobiles, is horsepower. One horsepower is equal to 746 watts.

Simple Machines

Machines are tools that make doing work easier. Instead of carrying those heavy moving boxes across the house and down the driveway, it would have been easier to put them on a dolly with wheels and roll them to the truck—in other words, use a simple machine to help do the work. There are six types of simple machines, and each one makes performing work easier in a different way.

Wheel and Axle

A **wheel and axle** is a simple machine consisting of two parts: a wheel and an axle, which is a long rod that fits through an opening in the wheel and allows the wheel to turn and move a load. Moving those heavy boxes through the house and down the driveway using a dolly is an example of a wheel and axle.

Lever

A **lever** is a rigid arm that pivots on a fulcrum, or fixed point, that helps to move or lift an object when pressure is applied to the other side. For example, the claw part of a hammer works to lift a nail out of a board when the hammer's handle is pushed down. Similarly, the blade of a shovel works to lift a heavy pile of dirt when the handle of the shovel is pushed down. Both of these are levers.

Inclined Plane

An **inclined plane** is a flat surface that is inclined, or slanted, and helps move objects along distances. Pushing those heavy moving boxes up a ramp into the truck is a lot easier than lifting them up and placing them into the truck yourself.

Wedge

Two inclined planes placed back-to-back come together to form a sharp edge called a **wedge**, used to push two objects apart. The blade on an axe and the blade on a knife are examples of wedges.

Screw

An inclined plane wrapped around a cylinder is a **screw**; the inclined plane is also called the thread of the screw, and it forms a sharp edge that can hold objects down, hold objects together, or work to crush through an object. Lightbulbs, jar lids, and fan blades are all examples of screws.

Pulley

A **pulley** consists of a rope that fits into a wheel, with one end of the rope attached to the load, or the object that is to be moved. When the rope of the pulley is pulled, the wheel turns and helps to move the load. Imagine you had to move a piano to the fifth floor of an apartment building. It would be much easier to use a pulley to lift the piano through a window than to carry the piano up several flights of stairs. Pulleys allow loads to be moved up, down, or sideways.

Two or more simple machines can be put together in different ways to form **complex machines**. For example, a wheelbarrow consists of a lever (the handles) and a wheel and axle (the wheels).

LESSON REVIEW

Complete the activities below to check your understanding of the lesson content. The Unit 3 Answer Key is on page 154.

Vocabulary

Fill in the blanks.

1. The unit of work is the ______________________.

2. A machine made up of two or more simple machines is a ______________________.

3. ______________________ is the capacity for doing work.

Skills Practice

Answer the questions based on the content covered in the lesson.

4. Which four simple machines make up a bicycle, and what are their jobs?

____

____

____

____

5. Carrie is trying to move a very heavy log from her yard. She pushes on the log for almost an hour, but it does not budge. She calls her friend Nina to help her move the log. They push on the log together, but it still does not move. Carrie calls another friend, Pia, and all three push on the log. Finally, it begins to move, and they roll it off the grass.

   Describe the work being done on the log when Carrie pushes on the log alone, when Nina and Carrie push on the log together, and when Carrie, Nina, and Pia push together. Be sure to explain your reasoning.

____

____

____

____

6. Matteo expends 60 joules of energy to move his stalled car off the road. How much work did Matteo do on the car? Assume no energy is lost as heat.

   A. 40 joules

   B. 50 joules

   C. 60 joules

   D. 70 joules

Key Terms

atom
compound
distillation
electrons
element
evaporation
filter
filtration
heterogeneous
homogeneous
mixture
molecule
neutron
nucleus
proton
pure substance

Structure of Matter

Just about everyone has seen copper metal, breathed oxygen gas, tasted salt, and poured a glass of milk. These are all different materials, but what makes them different? Is it what they are made of? Is it the way they are put together? Scientists can group them based on both what they are made of and how they are put together, as well as predict some of their other properties.

Atoms and Elements

Elements, like copper, are pure substances that consist entirely of the same type of **atoms**. An atom is the smallest particle into which an element can be broken down and still have the properties of that element. All copper atoms have the same properties, such as how they react with atoms of other elements.

Atoms have a central core, called a **nucleus**, which consists of smaller particles called **protons** and **neutrons**. Still smaller particles, called **electrons**, are found around the atom, outside the nucleus. Protons are positively charged, electrons are negatively charged, and neutrons are neutral. Copper has 29 protons, and a neutral copper atom will have 29 electrons to balance out the charges. Copper has the symbol Cu and is the 29th element on the periodic table. (There is an example of the periodic table in Lesson 10.)

Oxygen is also an element; it is the 8th element on the periodic table, so it has 8 protons and 8 electrons. Oxygen, as well as some other elements, is rarely seen as a single atom. It forms a **molecule**, made up of two atoms bonded together. The symbol for oxygen is O, but we use the symbol O_2 to represent the oxygen molecule that makes up oxygen gas.

Compounds

If atoms from two or more different elements bond together, they can form molecules, and the substance the molecules create is called a **compound**. Carbon dioxide is a compound. The symbol for carbon is C, and the symbol for oxygen is O. Carbon dioxide has the formula CO_2, which shows that a molecule has one carbon atom and two oxygen atoms. Salt is also a compound; it is made from chemically reacting sodium atoms with chlorine atoms. Each sodium atom must be matched by a chlorine atom, and we show this one-to-one relationship in the formula for salt, NaCl. Note that particles of NaCl are not molecules because the atoms bond in a special way.

Compounds and elements are **pure substances**. They are made entirely of identical atoms or molecules. If they contain more than one element, the elements are bonded together in a fixed ratio. The ratio can't be changed without forming another compound. For example, carbon monoxide (CO) is a very different gas from carbon dioxide, even though they both contain carbon and oxygen.

Classification of Matter

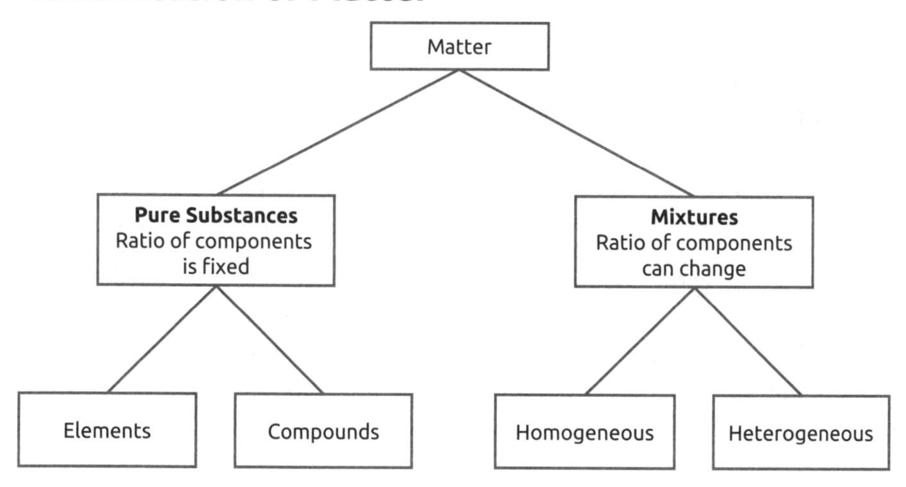

Mixtures

Salt water and milk are examples of **mixtures**. The ratio of the components of mixtures can change without changing the identity of the mixture. Whether you make salt water by adding a pinch or a handful of salt to a cup of water, it is still salt water. Salt water is an example of a **homogeneous** mixture. A homogeneous mixture looks uniform throughout and can be difficult to tell from a pure substance. Salt water and water look identical; you would need to test them to tell them apart.

Milk is an example of a **heterogeneous** mixture. It consists mostly of water, proteins, sugars, and fat. A heterogeneous mixture's components can be seen as different from each other. The protein molecules in milk are large enough to scatter light as it goes through the water, which is why milk looks white. The scattering makes it obvious there is something dissolved in the water. It is usually pretty easy to tell a heterogeneous mixture from a pure substance.

The ratio of components in milk can change, but the result will still be "milk." For example, the fat content can be lowered, so people have a choice of whole milk, 2% fat milk, 1% fat milk, and skim milk. When you are deciding whether something is a pure substance or a mixture, it can help to imagine changing the ratio of the components. If you end up just changing a "recipe" (such as changing the amount of fat in milk), it's a mixture. If you end up changing a chemical formula, and therefore making a different substance, it's a pure substance.

Helpful Hint

Pure substances have formulas, while mixtures have recipes.

Separating Mixtures

Another way to determine if a substance is a mixture is to try separating it into its components. Some heterogeneous mixtures consist of a solid in a liquid or a solid in a gas, such as sand in water or dust in air. These mixtures can be separated by **filtration**, a process in which the mixture is put through a **filter**. The filter allows one component to go through while trapping the other component. Coffee filters are paper filters, with holes small enough to trap the coffee grounds but large enough to let the water through.

Two liquids that are mixed together can be separated by **distillation** if they have different boiling points. Heating the mixture to the lower boiling point will evaporate one liquid and leave behind the liquid that has the higher boiling point. **Evaporation** also separates dissolved solids from liquids. When the mixture is heated, the liquid boils and evaporates, leaving the solid behind.

Unit 3 Lesson 8 LESSON REVIEW

Complete the activities below to check your understanding of the lesson content. The Unit 3 Answer Key is on page 154.

Skills Practice

Answer the questions based on the content covered in the lesson.

1. Describe the difference between pure substances and mixtures.

____

____

____

2. Describe the difference between homogenous mixtures and heterogeneous mixtures.

____

____

____

LESSON REVIEW

Select the best option.

3. Air at sea level is about 20% oxygen gas and 80% other gases. In hospitals, patients may be given air with a higher percentage of oxygen gas to help them breathe. Which kind of matter is air?

   A. element

   B. compound

   C. homogeneous mixture

   D. heterogeneous mixture

4. Sugar contains carbon, oxygen, and hydrogen atoms and has the formula $C_{12}H_{22}O_{11}$. Which kind of matter is sugar?

   A. element

   B. compound

   C. homogeneous mixture

   D. heterogeneous mixture

5. Argon is a gas made entirely of argon atoms. Argon is an example of (an element, a compound).

6. Gelatin is made by dissolving proteins in water, and it looks slightly cloudy in strong light. Gelatin is an example of a (homogeneous, heterogeneous) mixture.

7. Sugar water is clear; if you leave a cup of sugar water out, the water will evaporate, leaving the solid sugar. Sugar water is an example of a (pure substance, mixture).

8. When you cook spaghetti, you can separate it from the water in which it was cooked by draining it into a colander, which is a bowl with holes to let the water drain out. This is an example of separating a mixture by (filtration, distillation).

PROPERTIES OF MATTER

Key Terms

chemical change
chemical property
density
hardness
phase
phase change
physical change
physical properties
solubility

Properties and Changes

Everything has its own set of properties. For example, water dissolves many substances, while copper conducts electricity, and carbon dioxide puts out fires. These substances can also undergo changes. Some changes will affect their properties, and some will not. Scientists measure these properties and changes to identify unknown substances. They can also predict what properties and changes to expect in a new substance.

Physical and Chemical Properties

Physical properties are properties that can be observed without changing the chemical identity of the substance. For example, the property of **hardness** can be measured on a salt crystal by measuring the amount of force needed to crush the crystal. The crystal won't look the same afterward, but it is still salt. Another useful physical property to measure is **solubility**. The solubility of sugar in water indicates how much sugar can be dissolved in a certain amount of water at a certain temperature. The identity of the sugar does not change. If the water is removed by evaporation, what is left behind is still sugar. A **chemical property** can be observed only when a substance takes part in a chemical reaction. For example, a chemical property of sugar is that it reacts with oxygen when heated, producing carbon dioxide and water.

Physical and Chemical Changes

A **physical change** is one in which the chemical identity of the substance is not changed. A **phase change** is a physical change to the shape of a substance, such as crushing it or ripping it apart. A **chemical change** is one in which the chemical identity of the substance is changed. The physical properties will change as well. In the previous example, sugar is changed into carbon dioxide and water, two substances that have different properties from those of sugar.

Phase Changes

All matter exists in one of three **phases**: solid, liquid, or gas. For example, at –25°C, water is a solid, but at +25°C, it is a liquid. A solid has a definite shape, which does not change to fit its container. It also has a definite volume, which doesn't change as a result of pressure changing. Liquids have an indefinite shape, so they take the shape of the container holding them. Liquids have a definite volume, similar to solids, while a gas has both an indefinite shape and an indefinite volume. A gas will change shape to fit its container and will also expand or contract to fill the container.

The melting point of water, as it warms from a solid to a liquid, is 0°C. If the water were being cooled from a liquid to a solid, it would freeze at the same temperature. The boiling point of water, as it is warmed from a liquid to a gas, is 100°C. If the water were being cooled from a gas to a liquid, it would condense at the same temperature. Substances have characteristic melting and boiling points. For example, aluminum has a melting point of about 660°C and a boiling point of about 2519°C, while carbon dioxide has a melting point of about –78°C and a boiling point of about –56°C.

Melting points and boiling points are physical properties. They can be measured without changing the chemical identity of the substance. Melting and boiling are examples of phase changes. During melting or boiling, a substance is changing from one phase to another. Since temperatures of phase changes are different for every substance, they can be useful in identifying unknown substances.

Helpful Hint

Melting point and freezing point are the same for a substance; likewise, boiling point and condensation point are the same.

Density

Density is another physical property that is very useful in identifying substances. **Density** describes the ratio between the volume a sample of the substance takes up and the mass of that sample. We find density by measuring the volume and mass of a sample and using them in this equation:

$$density = \frac{mass}{volume} \qquad d = \frac{m}{v}$$

For example, if you had a block of copper that had a mass of 123g and a volume of 13.8 cm^3, you would divide 123 by 13.8 to get a density of approximately 8.91 g/cm^3. Pure water has a density of about 0.9982 g/cm^3. Anything that floats in water must have a lower density than 1 g/cm^3.

Unit 3 Lesson 9 LESSON REVIEW

Complete the activities below to check your understanding of the lesson content. The Unit 3 Answer Key is on page 154.

Skills Practice

Answer the questions based on the content covered in the lesson.

1. Describe the difference between physical properties and chemical properties.

____

____

2. Most metals are shiny. Explain why shininess is a physical property.

____

3. Is a phase change a chemical change or a physical change? Explain your answer.

____

____

LESSON REVIEW

Select the best answer choice.

4. Which of the following is a chemical change?

   A. dissolving salt in water

   B. measuring the density of wood

   C. exposing iron to humid air to form rust

   D. melting a piece of butter

5. Carbon dioxide has a melting point of about −78°C and a boiling point of about −56°C. At which temperature is carbon dioxide a liquid?

   A. −134°C

   B. −65°C

   C. −25°C

   D. +15°C

6. You have 11.2 cm^3 of mercury, and it has a mass of 151.2 g. What is the density of your mercury sample?

   A. 0.07 g/cm^3

   B. 13.5 g/cm^3

   C. 162.4 g/cm^3

   D. 1693.4 g/cm^3

Chemical Equations

During a **chemical reaction**, atoms separate from each other. They then rearrange to form new bonds. A chemical equation shows what happens during the chemical reaction. For example, if hydrogen gas and oxygen gas react together to form liquid water, the chemical equation would show:

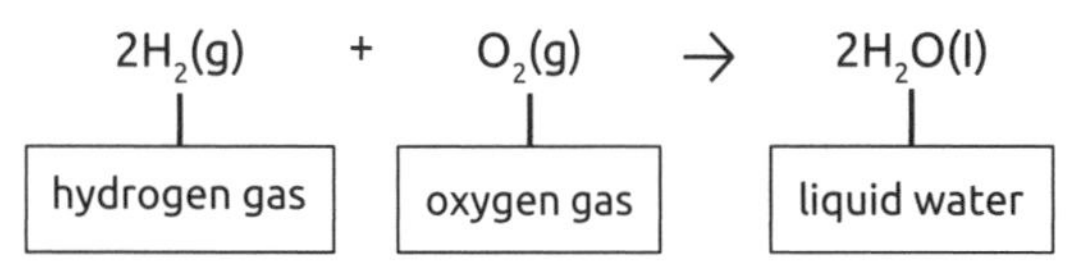

The symbols on the left side of the arrow represent the **reactants**. In this case, they are hydrogen gas and oxygen gas. The symbols on the right side of the arrow represent the **products**. In this case, the product is liquid water.

The symbols after the element/compound symbols indicate the state of matter: solid, liquid, or gas.

The numbers in front of the element/compound symbols indicate the ratios of the reactants and products to each other.

Put in words, the equation $2H_2(g) + O_2(g) \rightarrow 2H_2O(l)$ says: "Two molecules of hydrogen gas react with one molecule of oxygen gas to form two molecules of liquid water."

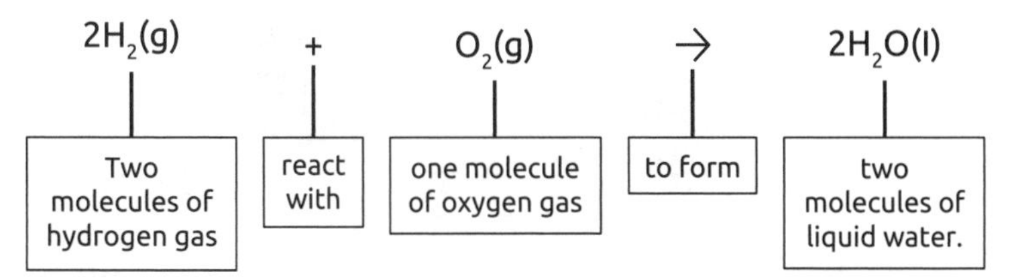

In this reaction, two elements reacted to form a compound. But compounds can also break down to form elements or other compounds. For example, solid sodium bicarbonate ($NaHCO_3$; baking soda) can react to form solid sodium carbonate (Na_2CO_3) plus carbon dioxide (CO_2) and water gases (H_2O):

$$2NaHCO_3(s) \rightarrow Na_2CO_3(s) + CO_2(g) + H_2O(g)$$

The first reaction (to form water) is called a synthesis reaction because a compound is being put together. The second reaction (to break down sodium bicarbonate) is called a decomposition reaction because a compound is being taken apart.

There are also reactions called substitution, or displacement reactions, in which compounds are both taken apart and put together, but in different arrangements. For example, aqueous (dissolved in water) sodium chloride (NaCl) reacts with aqueous silver nitrate ($AgNO_3$) to produce aqueous sodium nitrate ($NaNO_3$) and solid silver chloride (AgCl):

$$NaCl(aq) + AgNO_3(aq) \rightarrow NaNO_3(aq) + AgCl(s)$$

Balanced Equations

In the last equation, no numbers were in front of the compound symbols. This is because using one of each reactant and product produced a **balanced equation**. An equation is balanced

Key Terms

- atomic mass unit (amu)
- balanced equation
- chemical equation
- coefficient
- Law of Conservation of Matter
- limiting reactant
- periodic table
- products
- reactants

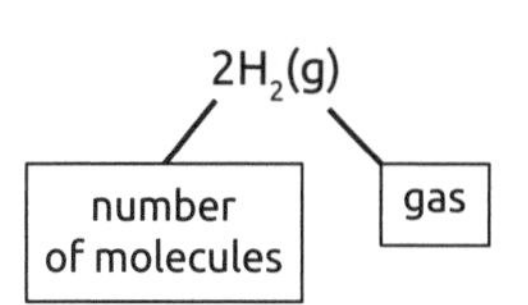

Skills Tip

The coefficient and the subscript are the numbers that serve different purposes in chemical equations. The coefficient appears before the chemical symbol and indicates the number of atoms of that element. If there is no coefficient, there is 1 atom.

The subscript is the small number and tells how many atoms of the element are in a compound. You will never see 1 as a subscript because you can assume that the element has at least 1 atom.

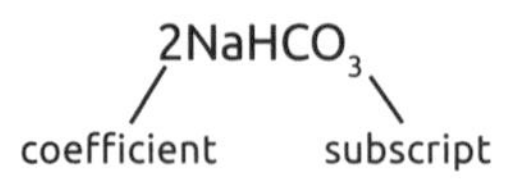

when the same number of each type of atom appears on both sides. You can use the chemical symbols for each type of atom in the above equation and count the number of atoms:

oxygen O
sodium Na
chlorine Cl
silver Ag
nitrogen N

You can look for the symbols in the equation and count on each side of the equation:

$NaCl(aq) + AgNO_3(aq) \rightarrow NaNO_3(aq) + AgCl(s)$

You will find that there are 7 of each type of atom on each side of the equation.

If the equation isn't balanced with one of each reactant and product, **coefficients** must be added to make it balanced. It can help to use a table to keep track of the numbers and types of atoms.

$2NaHCO_3(s) \rightarrow$				$Na_2CO_3(s) + CO_2(g) + H_2O(g)$			
Reactants				**Products**			
Na	H	C	O	Na	H	C	O
2 from 2 $NaHCO_3$	2 from 2 $NaHCO_3$	2 from 2 $NaHCO_3$	6 from 2 $NaHCO_3$	2 from 1 Na_2CO_3	2 from 1 H_2O	2 from 1 Na_2CO_3 and 1 CO_2	6 from 1 Na_2CO_3 and 1 CO_2 and 1 H_2O
2	2	2	6	2	2	2	6

Notice that we multiplied the coefficient by the subscript for oxygen. For each $NaHCO_3$, there are 3 oxygen atoms. This means that for every 2 $NaHCO_3$, there are 6 oxygen atoms. This is because the coefficient (2) means "two times" each element and its atoms in the compound. In the balanced equation, there are 12 atoms on the reactants side and 12 atoms on the products side. The number of sodium, hydrogen, carbon, and oxygen atoms do not change, although the way they are arranged does. During a chemical reaction, the chemical identity of the reactants changes, but the mass stays constant. This is because, according to the **Law of Conservation of Matter**, matter cannot be created or destroyed.

The equation for the reaction of propane with oxygen gas is not balanced when one of each reactant and product is used:

$C_3H_8(g) + O_2(g) \rightarrow CO_2(g) + H_2O(g)$

$C_3H_8(g) + O_2(g) \rightarrow$			$CO_2(g) + H_2O(g)$		
Reactants			**Products**		
C	H	O	C	H	O
6 from C_3H_8	8 from C_3H_8	2 from O_2	1 from CO_2	2 from H_2O	2 from CO_2 and 1 from H_2O
3	8	2	1	2	3

If we add the coefficient 3 to CO_2, that will balance the number of carbon atoms, but it will change the number of oxygen atoms on the product side to 7 (6 from CO_2 and 1 from H_2O). If we then add the coefficient 4 to H_2O, it will balance the number of hydrogen atoms, but it will change the oxygen atoms again. Now there are 10 oxygen atoms on the product side (6 from $3CO_2$ and 4 from $4H_2O$). The equation needs to have 10 oxygen atoms on the reactant side to balance. We can achieve that by adding the coefficient 5 to the O_2.

$C_3H_8(g) + 5O_2(g) \rightarrow$			$3CO_2(g) + 4H_2O(g)$		
Reactants			**Products**		
C	H	O	C	H	O
3 from C_3H_8	8 from C_3H_8	10 from $5O_2$	3 from $3CO_2$	8 from $4H_2O$	6 from $3CO_2$ and 4 from $4H_2O$
3	8	10	3	8	10

Information from the Periodic Table

Chemists do not count out individual atoms to make sure the reactions they are doing have the right proportion of reactants. Instead, they measure the mass of the reactants. They can then use the mass of each element to determine the number of atoms and molecules. Information about the mass of each element is listed in the **periodic table** of the elements.

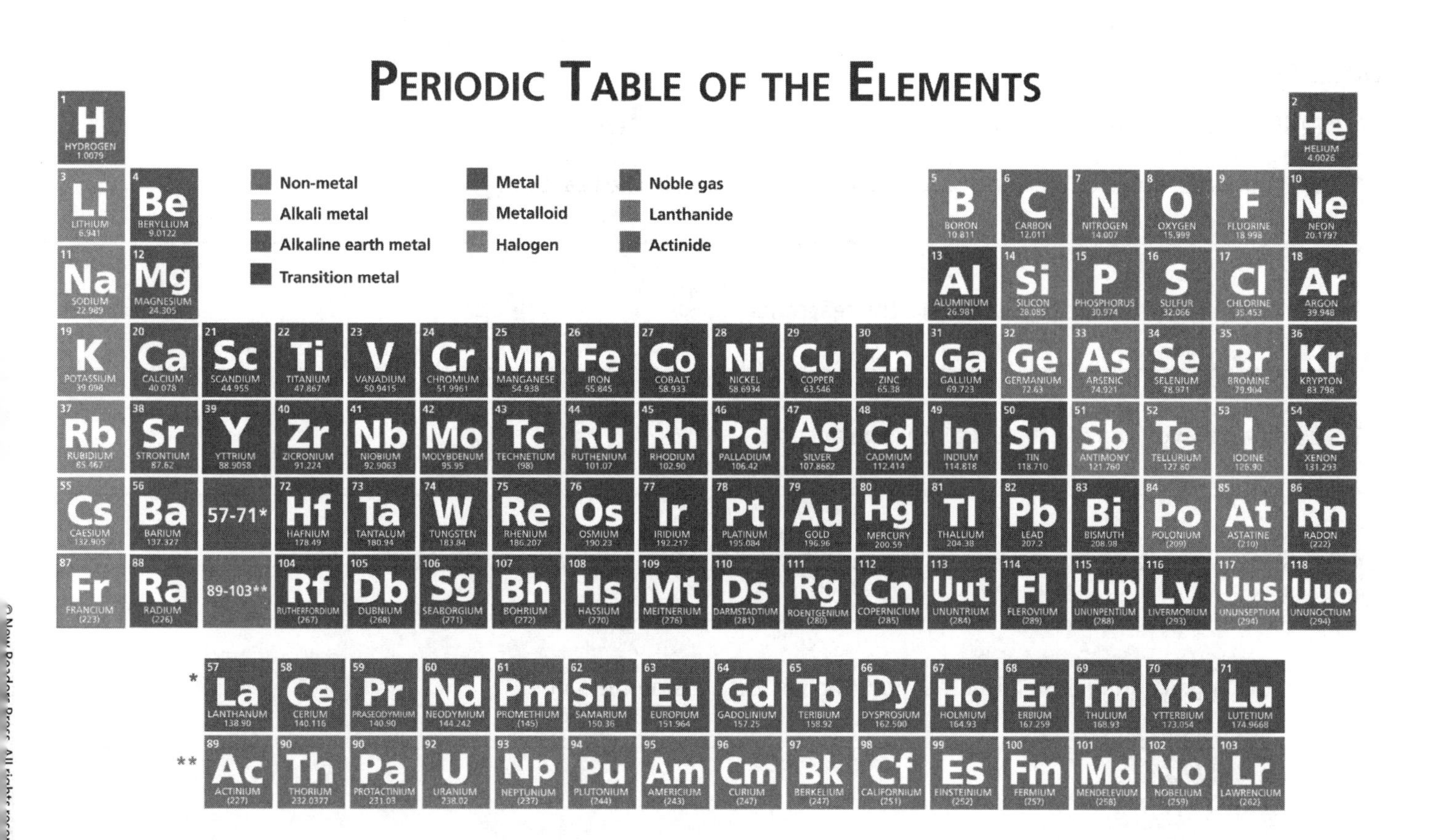

The numbers under each element symbol indicate the average mass for an atom of that element, in **atomic mass units** (amu). If there are two numbers, it indicates the range for that element, depending on the source. An amu is very small; it is much easier to measure mass in grams than in amus. For example, according to the table, 19.00 grams of fluorine will have the same number of atoms as 22.99 grams of sodium.

Limiting Reactant

If the amounts of each reactant do not match the amounts in the balanced equation, one of the reactants will be used up before the rest. At that point, the reaction will end, even if there are other reactants left over. The reactant that is used up first is called the **limiting reactant**. For example, in the reaction of propane and oxygen, if the actual ratio of oxygen to propane is 10:1 instead of 5:1, the reaction will end when all of the propane is used up, leaving half of the oxygen unreacted. In that reaction, propane is the limiting reactant.

Unit 3 Lesson 10 LESSON REVIEW

Complete the activities below to check your understanding of the lesson content. The Unit 3 Answer Key is on page 154.

Skills Practice

Answer the questions based on the content covered in the lesson.

Use the following reaction to answer questions 1–5.

$$2Al(s) + 3CuCl_2(aq) \rightarrow 2AlCl_3(aq) + 3Cu(s)$$

1. What are the states of matter of each element/compound in the reaction?

____

2. Which are the products and which are the reactants?

____

3. Explain how you can tell the equation is balanced.

____

4. Restate the equation in words; use “copper chloride” for $CuCl_2$ and “aluminum chloride” for $AlCl_3$.

____

5. What changes during a chemical reaction, and what must stay the same?

____

LESSON REVIEW

Use the following reaction and partial table to answer questions 6–10.

$C_5H_{12}(g) + O_2(g) \rightarrow CO_2(g) + H_2O(g)$

$C_5H_{12}(g) + O_2(g) \rightarrow$			$CO_2(g) + H_2O(g)$		
Reactants			**Products**		
C	H	O	C	H	O
___ from C_5H_{12}	___ from C_5H_{12}	___ from O_2	___ from CO_2	___ from H_2O	___ from CO_2 and ___ from H_2O
___	___	___	___	___	___

6. In the unbalanced equation, how many carbon atoms are on each side?

____

7. In the unbalanced equation, how many hydrogen atoms are on each side?

____

8. In the unbalanced equation, how many oxygen atoms are on each side?

____

9. In the unbalanced equation, how many total atoms are on each side?

____

10. Which set of coefficients will balance the equation?

    A. 5, 2, 2, 1

    B. 1, 5, 5, 6

    C. 1, 2, 5, 8

    D. 1, 8, 5, 6

Select the best option.

11. In the reaction $NaCl(aq) + AgNO_3(aq) \rightarrow NaNO_3(aq) + AgCl(s)$, if the actual ratio of NaCl to $AgNO_3$ is 1.5:1, then (NaCl, $AgNO_3$) will be the limiting reactant.

12. The reaction $2NaHCO_3(s) \rightarrow Na_2CO_3(s) + CO_2(g) + H_2O(g)$ is balanced because the total number of (atoms, compounds) is the same on each side.

13. According to the periodic table, the mass of an atom of iron (Fe) is about (26, 55.85) amu.

14. According to the periodic table, the number of atoms in 39.10 grams of potassium (K) will be the same as the number of atoms in (39.10, 40.08) grams of calcium (Ca).

SOLUTIONS

Key Terms

acid
base
concentrated
dilute
neutral
pH
precipitate
saturated
solubility
solute
solvent
strong acid
strong base
supersaturated
unsaturated
weak acid
weak base

All living things depend on solutions. Blood cells float in salt solutions that help them keep their shape. Our digestive systems use solutions to keep digestive enzymes active. The oceans, which are the largest solution in the world, are home to most of the life on the planet.

Solubility

Solubility of a substance is a physical property. It indicates how much of a substance can be dissolved. It depends a great deal on temperature. It also depends on the **solvent,** which is what the substance is being dissolved in. The substance being dissolved is known as the **solute.** A **concentrated** solution has a lot of solute. A **dilute** solution has a small amount of solute. Solubility is often expressed as the number of grams of solute that can be dissolved in 100 grams of solvent, at a particular temperature. It is helpful to use a solubility curve to predict the solubility of different solutes at different temperatures.

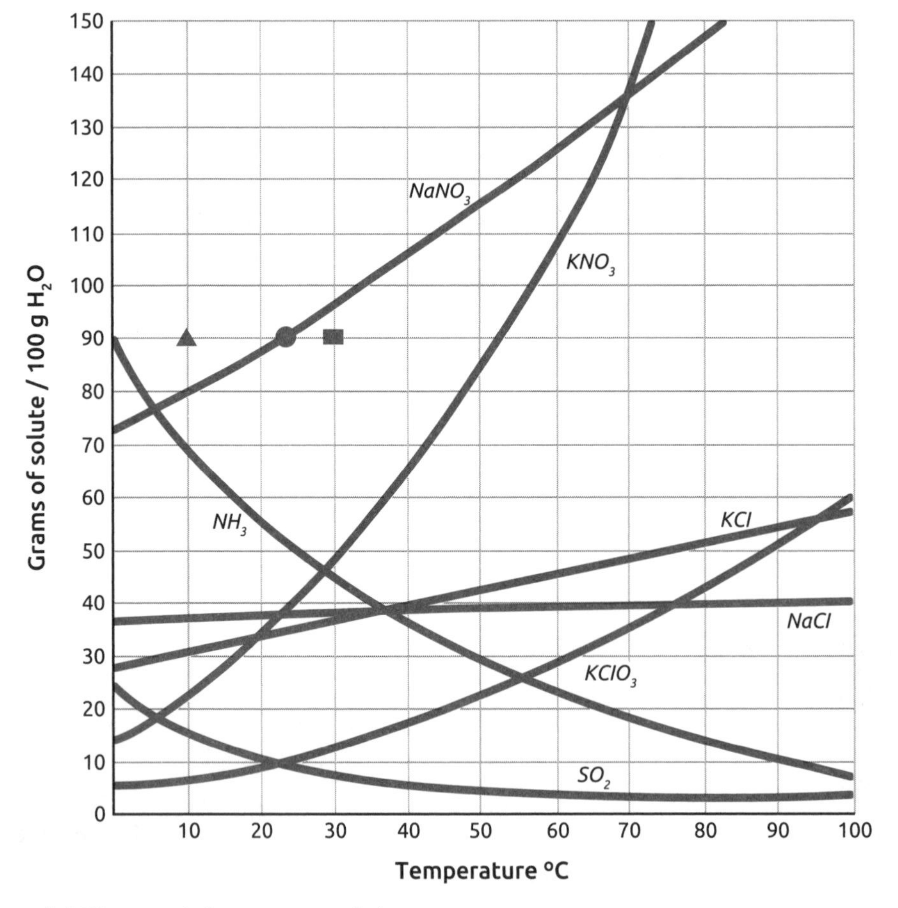

Solubility graph for compounds in water

The lines for each substance indicate the most solute that will dissolve in 100 grams of water at any particular temperature. For example, at about 24°C, you can dissolve 90 grams of $NaNO_3$ in 100 grams of water (point is marked with a circle). Solubility increases with temperature for most solids and decreases with temperature for most gases. Both NH_3 and SO_2 are gases. Notice that at 30°C, 90 grams is below the $NaNO_3$ line (point is marked with a square). At 10°C, 90 grams is above the $NaNO_3$ line (point is marked with a triangle).

The circle, or any point on the line, indicates a solution that is **saturated** at that temperature. If you add more solute to a saturated solution, it will not dissolve. The square, or any point below the line, indicates an **unsaturated** solution at that temperature. If you add more solute to an unsaturated solution, it will dissolve. The triangle, or any point above the line, indicates a **supersaturated** solution. This means you have dissolved more solute than can normally be dissolved. You can only do this by making a saturated solution at one temperature and then cooling the solution. Supersaturated solutions are not stable, and solute will start to **precipitate**. This means it will "undissolve" until the solution is no longer supersaturated.

Acid and Base Solutions

Some solutes, when dissolved in water, form an **acid** or **base** solution. Acids form positive hydrogen ions (H^+) in water, while bases form negative hydroxide ions (OH^-) in water. Some acids form a lot of hydrogen ions for every gram of solute. These are called **strong acids**. Those that form only a small amount of hydrogen ions for the same amount of solute are called **weak acids**.

We use a **pH** scale to indicate how acidic or basic a solution is. The scale goes from 0 to 14. A pH measurement of less than 7 means the solution is acidic. A pH measurement of 7 means the solution is **neutral**. Pure water has a pH of 7. The lower the number, the more acidic the solution is. A strong acid will lower the pH of water more than a weak acid will. A pH of more than 7 means the solution is basic. A **strong base** will raise the pH of water more than a **weak base** will.

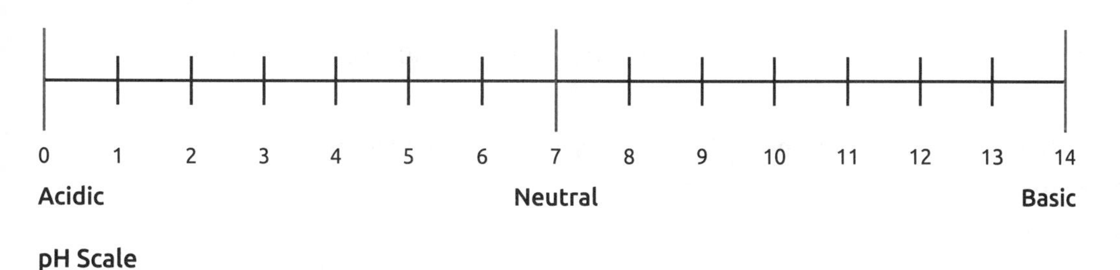

pH Scale

Real-World Connection

Rock candy is just sugar that has precipitated from a supersaturated solution. Dissolve as much sugar as you can in boiling water. Let it cool a bit, and then transfer it to a jar. Hang a clean string from the lid into the solution. During the next few weeks, sugar will precipitate and form crystals on the string.

LESSON REVIEW

Complete the activities below to check your understanding of the lesson content. The Unit 3 Answer Key is on page 155.

Skills Practice

Answer the questions based on the content covered in the lesson.

1. Describe the difference between saturated, unsaturated, and supersaturated solutions.

____

____

____

2. Describe the difference between an acid and a base.

____

____

Select the best option.

3. Based on the solubility curve shown on page 108, which would a solution made from 70 grams of KNO_3 at 60°C be?

   A. saturated

   B. unsaturated

   C. supersaturated

4. Based on the solubility curve shown on page 108, which would a solution made from 70 grams of KCl at 60°C be?

   A. saturated

   B. unsaturated

   C. supersaturated

5. Based on the solubility curve shown on page 108, about (30, 50) grams of NH_3 in 100 grams of water at (30°C, 50°C) would make a saturated solution.
6. Acetic acid, the acid in vinegar, is a weak acid. This means it will change the pH of water (more, less) than the same amount of a strong acid will.
7. Lemon juice has a pH of about 2.5, which means lemon juice is (an acid, a base).
8. Chlorine bleach has a pH of about 12.8, which means chlorine bleach is (an acid, a base).

UNIT PRACTICE TEST

Answer the questions based on the content from this unit. The Unit 3 Answer Key is on page 155.

Select the best option.

1. Which has the least kinetic energy?

   A. a 30 kg car moving at 1 meter per hour

   B. a 30 kg car moving at 5 meters per hour

   C. a 50 kg car moving at 5 meters per hour

   D. a 50 kg car moving at 10 meters per hour

2. Compared to a rock sitting 10 feet off the ground, a rock that is 20 feet off the ground has (more, less) gravitational potential energy.

3. The energy a fox needs to run comes from the food it eats. This is an example of which energy transformation?

   A. kinetic to potential

   B. potential to chemical

   C. mechanical to kinetic

   D. chemical to mechanical

Fill in the blank.

4. An issue with ________________________ power plants is that they would be unreliable in a drought.

5. In ________________________ power plants, fission reactions release large amounts of energy.

6. When you hold a hot cup of coffee, your hands are warmed through the heat-transfer process of ________________________.

Select the best option.

7. Hand warmers use a chemical reaction to produce heat. This reaction is (endothermic, exothermic).

Base your answers to questions 8–10 on the following passage and the content in this unit.

X-rays are a type of radiation with a very high frequency and a short wavelength. The electromagnetic spectrum begins with radio waves, which have a low frequency and long wavelengths. Gamma rays are on the opposite end of the spectrum with even higher frequency and shorter wavelengths than x-rays.

X-rays are able to pass through low-density substances such as skin and soft tissue but are stopped by bones, teeth, and metal. Medical professionals use x-ray machines to obtain a clear picture of the hard structures inside a person. The rays are beamed through a person onto a piece of film; x-rays are not felt or seen by the patient. A doctor is able to use these x-ray pictures to determine if a bone has been broken or has any damage.

Security checkpoints also use x-ray machines to keep facilities safe and free of dangerous objects. At airports and judicial buildings, people are asked to remove any metal objects before they walk through an x-ray machine. The machine scans the person and detects if they have any metal on them. At airports, suitcases and all baggage must be scanned by an x-ray to be sure that no weapons or forbidden objects are present.

X-rays can cause damage to cells and tissues, so it is important to limit exposure to them. Often, when doctors or dentists take x-rays, they provide the patient with a lead vest. This lets the patient cover the areas that do not need to receive x-rays. The person who is actually taking the x-ray stands in a shielded area so that he does not expose himself to these helpful but dangerous rays.

8. Which statement describes the relationship between the frequency and wavelength of the electromagnetic spectrum?

   A. Radiation with a low frequency has a long wavelength.

   B. Radiation with a high frequency has a long wavelength.

   C. Radiation with a medium frequency has a very long wavelength.

   D. Radiation with a medium frequency has a very short wavelength.

9. Which statement about x-rays is true?
   - **A.** X-rays can be felt and seen by humans.
   - **B.** X-rays can pass through metal and bone.
   - **C.** X-rays have a longer wavelength than gamma rays.
   - **D.** X-rays are always dangerous and should never be used by people.

10. Order the following parts of the electromagnetic spectrum in order from longest wavelength (1) to shortest wavelength (3).

_____	gamma rays
_____	x-rays
_____	radio waves

Base your answers to questions 11–13 on the following passage and the content in this unit.

The checkered flag is waved, and the cars are off. A red car, blue car, and yellow car are racing on a course. The cars have the same mass. The red car is first, the yellow car is second, and the blue car is in last place. But wait—the blue car is preparing to pass the red and yellow cars. The driver's speed increases from 40 m/s to 52 m/s in 3.5 seconds. He passes both cars and stays east toward the finish line. The blue car finishes the race first in 25 seconds.

The racetrack is 800 meters long. The table shows the data from the racecars.

Data from the Racecars	
Car	**Time Race Completed**
red	29 seconds
yellow	35 seconds
blue	25 seconds

11. Which car had the highest average velocity? Explain your answer.

____

____

12. Which correctly describes the acceleration of the blue car when it went from 40 m/s to 52 m/s?
   - **A.** 3.4 m/s^2
   - **B.** 4.0 m/s^2
   - **C.** 12 m/s^2
   - **D.** 20.6 m/s^2

13. Five seconds into the race, the cars were traveling at these velocities: red: 27.6 m/s east, yellow: 22.9 m/s east, blue: 26.7 m/s east. Which car had the greatest momentum at that time?
   - **A.** red
   - **B.** yellow
   - **C.** blue

Base your answers to questions 14–16 on the following passage and the content in this unit.

Karissa and her team are building a house. They need to move larger bundles of lumber from the first story to the second story to construct the subfloor on that level. First, Karissa tries to lift the bundles on her own, but they do not move. Next, two members of her team join in, but the bundles still do not move. Two more team members join in, and all five of them are able to lift the bundle, but they have difficulty getting to the second story since the stairs have not been built yet.

14. When was work being done on the bundle of lumber? Explain your answer.

____

____

15. Karissa and her team do 2,000 joules of work to move the bundle of lumber. About how much energy does the team expend?
   - **A.** 300 joules
   - **B.** 400 joules
   - **C.** 1,000 joules
   - **D.** 2,000 joules

16. Which simple machine could Karissa's team use to move the bundle of lumber to the second story?
   - **A.** lever
   - **B.** pulley
   - **C.** screw
   - **D.** wedge

17. Matt stands on a skateboard and pushes on the wall with a force of 20 Newtons. Which statement best describes what happens next?

    A. Matt does not move because the forces between Matt and the wall are balanced.

    B. Matt will move from the wall because the wall pushes back with an equal but opposite force.

    C. Matt does not move because he is pushing on the wall but the wall is not pushing on him.

    D. Matt will move away from the wall because his inertia is greater than the wall's inertia.

18. Which simple machines work together to make this can opener?

    A. inclined plane, screw, wedge

    B. lever, wheel and axle, inclined plane

    C. wheel and axle, wedge, lever

    D. screw, pulley, lever

Use the following scenario and the content in this unit to answer questions 19–21.

A child has a toy car with a mass of 0.5 kg and a toy truck with a mass of 2kg. She pushes the toy truck across the floor, and it hits the toy car. The toy car then begins to move.

19. Which statement describes the forces on both toys when they hit each other?

    A. The force on the toy truck is greater because it has more mass.

    B. The force on both toys is equal because of Newton's Third Law.

    C. The force on both toys is equal because gravity acts on both toys.

    D. The force on the toy car is greater because it moves with a greater acceleration.

20. Is work being done on one toy or both toys? Explain how you know.

    ____

    ____

21. When the two toys collide, which will move with a greater acceleration? Explain your reasoning.

    ____

    ____

Base your answers to questions 22–25 on the following passage and the content in this unit.

Angela and Thomas spent a day at the beach. Near the shore, the water had a lot of sand in it from the beach. They collected some of the ocean water in a cup. Angela poured the contents of the cup through a paper towel into another cup. The sand stayed on the towel and the rest passed through the towel into the second cup. The liquid in the second cup was clear; it looked like pure water. Thomas set the second cup out in the sun. After a while, all that was left was a white solid material in the cup.

22. Which type of matter was in the first cup?

    A. element

    B. compound

    C. homogeneous mixture

    D. heterogeneous mixture

23. When Angela poured the contents of the first cup through the towel, she was separating the components by (filtration, distillation).

24. Which type of matter was in the second cup, before it was set out in the sun?

A. element

B. compound

C. homogeneous mixture

D. heterogeneous mixture

25. The white solid that was left in the cup contains salt, NaCl, which is a (pure substance, compound).

Select the best option.

26. An atom of sulfur has 16 protons. How many electrons does a neutral atom of sulfur have?

A. 8

B. 18

C. 16

D. 32

27. The charge on an electron is (+, –, 0), and the charge on a neutron is (+, –, 0).

Base your answers to questions 28–30 on the following passage and the content in this unit.

A group of chemistry students were given a piece of an unknown metal and were asked to identify it. They first heated it up and found the melting point was about 63°C. They were not able to boil it with their classroom equipment. Their teacher told them the boiling point was about 769°C. They cooled the metal back down into a solid block and made some measurements. The mass of the block was 3.63 grams, and the volume was 4.08 cm^3. When they put a sample of the metal in water, it produced a gas and a solid. Their teacher told them the solid had a melting point of about 462°C. They determined from the information that the unknown metal was lithium.

28. At 105°C, lithium is a (solid, liquid, gas).

29. The density of lithium, rounded to two decimal places, is ____________ g/cm^3.

30. The type of change that lithium goes through when added to water is

A. physical because the gas is produced as the result of a phase change.

B. chemical because the lithium was dissolved in water.

C. physical because the density of the solid was the same as lithium.

D. chemical because the gas and the solid had different properties from lithium.

Select the best option.

31. Melting and boiling are examples of (chemical, physical) changes.

32. One way to test for brittleness of a material is to measure how far a thin piece of the material can be bent before breaking. Brittleness is a (chemical, physical) property.

33. Potassium metal reacts with chlorine gas to make potassium chloride, which is used as a salt substitute. Potassium's ability to react with chlorine gas is a (chemical, physical) property.

Select the best option.

34. The Law of Conservation of Matter requires that the number of (atoms, compounds) be the same on both sides of the arrow in a chemical equation.

35. The average atomic mass of potassium is 39.1 amu, while of iodine it is 126.9 amu. This means that the number of atoms in 126.9 grams of iodine should be (about 3 times as many, exactly the same) as the number of atoms in 39.1 grams of potassium.

Use the following graph and the content in this unit to answer questions 36–38.

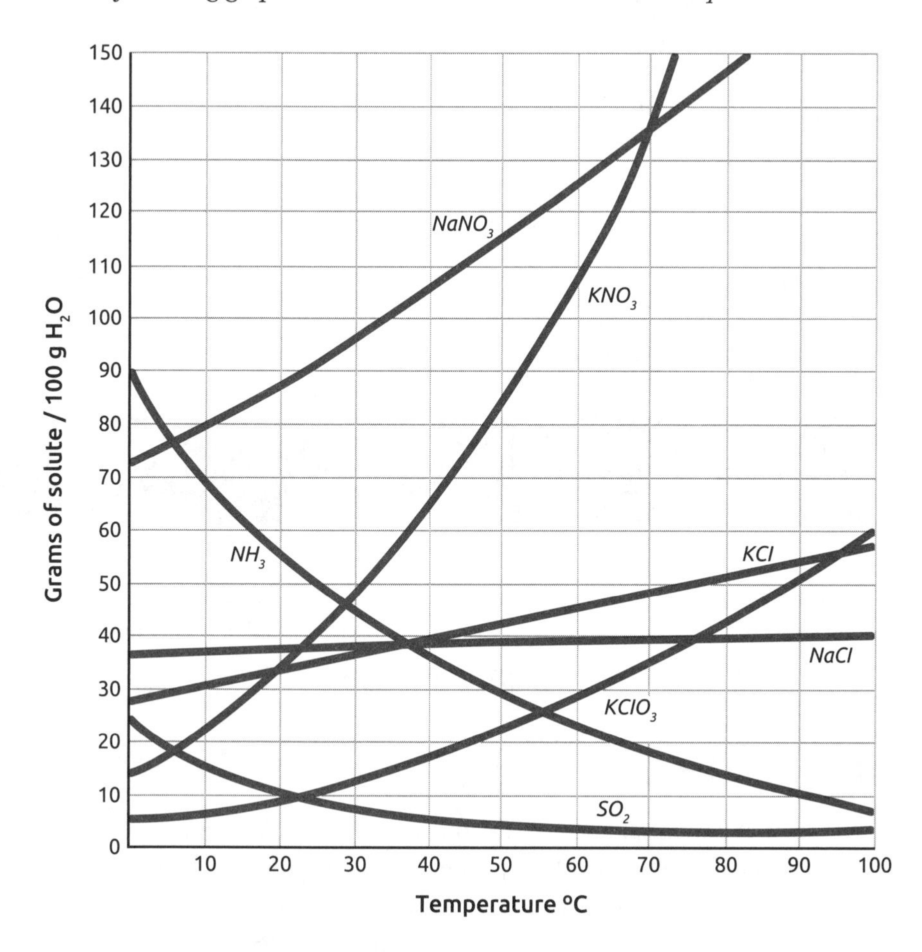

36. Which of the following would produce an unsaturated solution in 100 grams of water at 40°C?

 A. 30 grams of $KClO_3$

 B. 90 grams of KNO_3

 C. 50 grams of KCl

 D. 100 grams of $NaNO_3$

37. If you dissolve 70 grams of NH_3 in 100 grams of water at 10°C and then heat the solution to 30°C, you will have a (saturated, supersaturated) solution.

38. At temperatures below 70°C, $NaNO_3$ has a (lower, higher) solubility than KNO_3.

Select the best option.

39. Susan added 1 teaspoon of sugar to her iced tea while Clara added 2 teaspoons of sugar to the same amount of iced tea. Clara's iced tea is more (concentrated, diluted) than Susan's.

40. A chemist added an unknown substance to water; the pH of the solution was 9, indicating that the unknown substance was (an acid, a base).

41. When the same amounts of hydrochloric acid and citric acids are added to water, the hydrochloric acid solution has a lower pH than the citric acid solution. This indicates that hydrochloric acid is (stronger, more concentrated) than citric acid.

Earth Science

On a hike during a beautiful, sunny day, you stop for lunch near the bank of a river. You can see large rocks embedded in the water and a mountain rising up in the distance, creating a picturesque view. As the sun sets, the moon and the stars light up the night sky, but they are quickly covered by clouds as a steady rain begins to fall.

The atmosphere, land, and water are all interconnected. Earth and space science explores these connections and Earth's place in the universe. This branch of science helps us understand our planet and the universe beyond.

Unit 4 Lesson 1 MATTER CYCLES

Key Terms

carbon cycle

denitrification

fossil fuels

natural resources

nitrogen cycle

Matter flows throughout Earth in a series of cycles. Never still, the planet is constantly changing, and these cycles exist to keep Earth and its atmosphere in balance. Human activity can interfere with these cycles, affecting our climate and life on Earth.

The Carbon Cycle

Carbon is the fourth most abundant element in the universe. Most of it is found in rocks, but it also exists in the atmosphere, ocean, plants, animals, soil, and fossil fuels. Carbon moves between these locations in a process called the **carbon cycle**. This cycle keeps the carbon on Earth in balance, preventing all the carbon from leaving one location and entering another.

Too much carbon (in the form of CO_2) in the atmosphere would cause temperatures to rise. The carbon cycle acts like a thermostat, keeping Earth's temperature stable.

This diagram demonstrates the path of carbon in the carbon cycle.

The Carbon Cycle

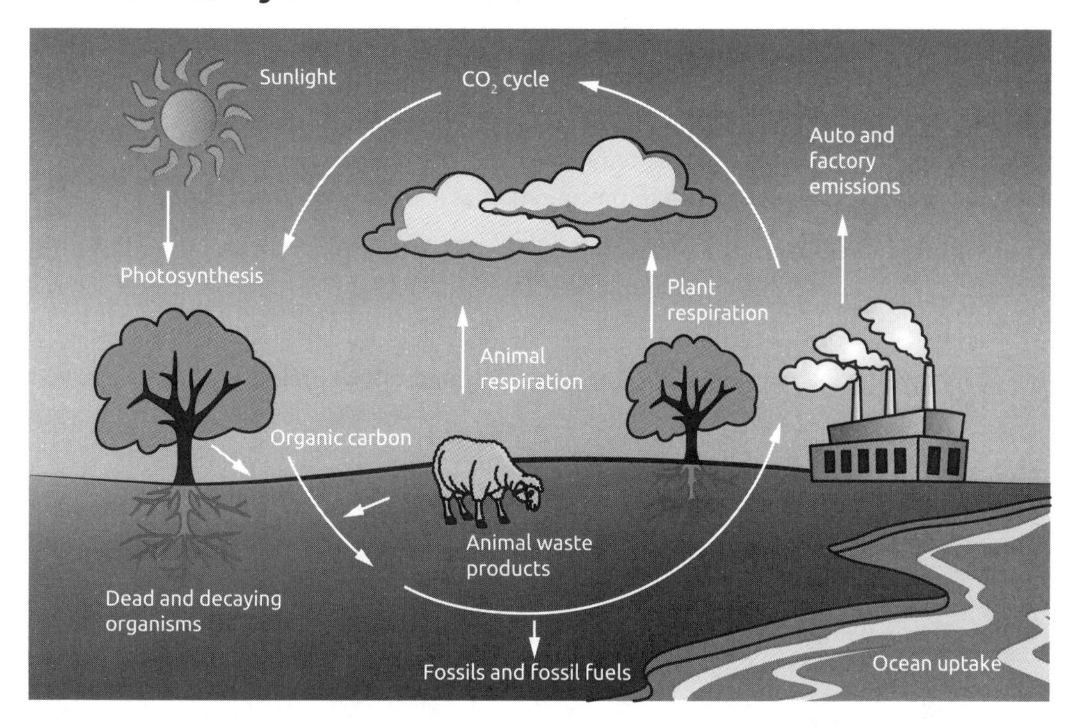

Natural Resources

Millions of years ago, organisms containing carbon died. Slowly, over time, this carbon became trapped and formed **natural resources** such as coal and oil. These are the **fossil fuels** we use today to run our cars and heat our homes. Carbon can be released back into the atmosphere through natural processes such as erosion and volcanic eruptions. It can also be returned through the burning of fossil fuels. The burning of fossil fuels, however, not only threatens to disrupt the balance of the carbon cycle, but it also uses a nonrenewable resource. Our fossil fuels are the remains of organisms that lived millions of years ago. Once they are used up, it will take millions of years for them to form again. Conservation of these types of resources will not only allow us to use them longer, but it will also help protect the balance of the natural cycles on Earth.

The Nitrogen Cycle

Nitrogen makes up almost 80 percent of the gases in Earth's atmosphere. Like carbon, it moves through Earth's systems in a process called the **nitrogen cycle**. Nitrogen is an essential element for plants and animals, and it can enter these organisms in several ways.

In one method, bacteria in the soil "fix" the nitrogen in the air to form nitrates that are taken up by plants. They are then passed to animals. Blue-green algae in the oceans can also fix nitrogen, making it available to organisms in the ocean. Finally, lightning fixes nitrogen so that it can fall to earth in precipitation and deposit nitrates into the soil.

When plants and animals die and decompose, the nitrogen is put back into the soil. Some bacteria carry out a process called **denitrification**, in which nitrates are converted back to nitrogen gas.

This diagram demonstrates the path of nitrogen in the nitrogen cycle.

Real-World Connection

Some human activities can interfere with natural cycles. The burning of fossil fuels adds more carbon into the atmosphere. Cutting down trees increases the amount of CO_2 in the air in two ways: When trees are burned, CO_2 is released. Having fewer trees also reduces the amount of CO_2 being removed from the atmosphere. The increase of carbon can raise Earth's temperature. Using nitrogen-based fertilizers on lawns and crops can add more nitrogen into the soil, rivers, and lakes, which can favor algae and plant growth, disrupting the ecosystem.

The Nitrogen Cycle

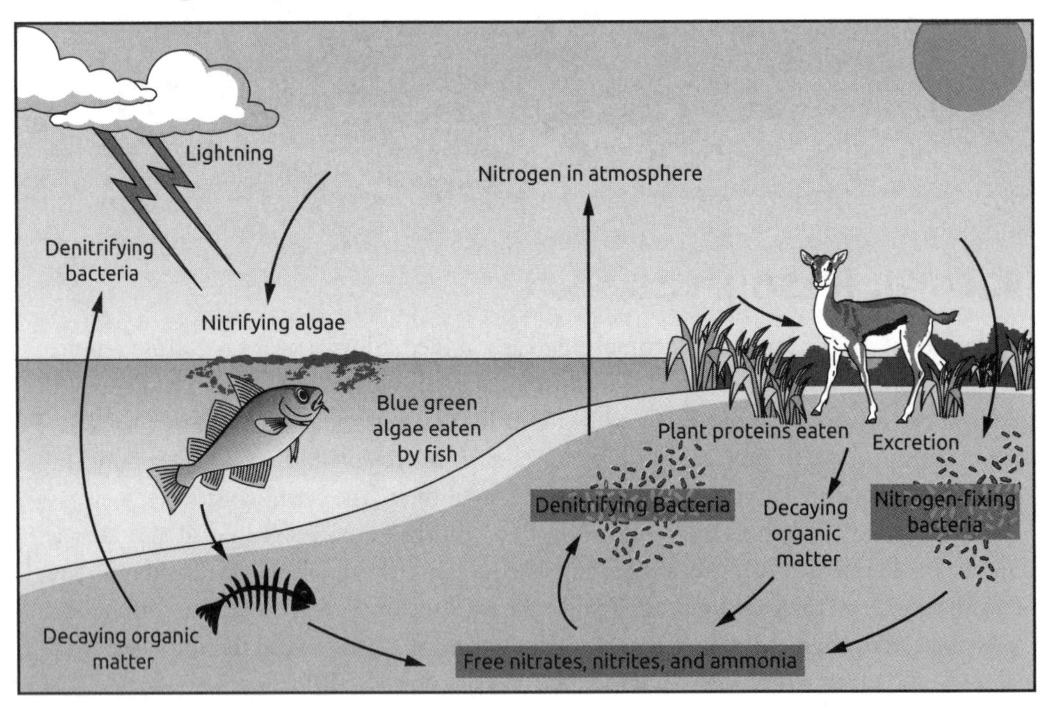

LESSON REVIEW

Complete the activities below to check your understanding of the lesson content. The Unit 4 Answer Key is on page 155.

Vocabulary

Write definitions in your own words for each of the key terms.

1. carbon cycle ______________________________

2. denitrification ______________________________

3. fossil fuels ______________________________

4. natural resources ______________________________

5. nitrogen cycle ______________________________

Skills Practice

Answer the questions based on the content covered in the lesson.

6. Each of these can fix nitrogen into nitrates EXCEPT which one?

   A. blue-green algae

   B. lightning

   C. denitrifying bacteria

   D. nitrogen-fixing bacteria

7. Which process releases carbon back into the atmosphere?

   A. lightning strikes

   B. volcanic eruptions

   C. decay of organisms

   D. formation of fossil fuels

8. Which of these is a nonrenewable resource?

   A. bacteria

   B. carbon

   C. nitrogen

   D. oil

CHARACTERISTICS OF THE ATMOSPHERE AND OCEANS

Key Terms

Coriolis effect

current

El Niño

gyre

In 2009, the southern United States had increased rainfall. At the same time, Australia was facing a drought. Both continents were affected by **El Niño**. This is an uncommon condition caused by unusually warm temperatures in the tropical Pacific Ocean. However, the wind and water currents are always important. They affect the weather across the planet every day.

How Does Wind Form?

The sun beats down on Earth, heating the surface of the planet. However, not all areas receive equal amounts of the sun's warmth. At the Equator, the sun is almost always directly overhead, warming the air. At the poles, however, the air is much cooler. The warm air at the Equator rises and moves toward the poles, where the cooler air sinks and moves back toward the Equator. But, while the air is moving, Earth is also rotating. This causes winds to seem to move clockwise in the Northern Hemisphere and counterclockwise in the Southern Hemisphere. This apparent change in the motion of the wind is called the **Coriolis effect**.

The uneven heating of Earth's surface causes global wind systems called trade winds, westerlies, and easterlies. Near the Equator is a region called the doldrums, named for the light winds that blow there.

The Coriolis Effect

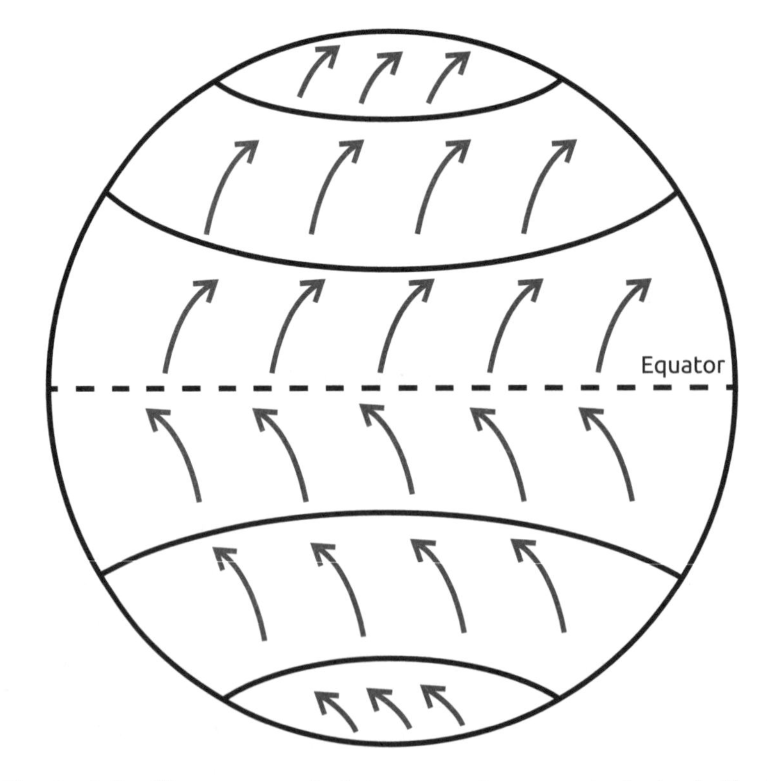

The Coriolis effect causes winds to appear to move clockwise in the Northern Hemisphere and counterclockwise in the Southern Hemisphere.

Wind and Water

The global winds blow across the surface of the ocean. They make the water move in the same direction the wind is blowing. Just like the wind, the ocean is also affected by the Coriolis effect. The Coriolis effect causes the water to form **currents** that move in spirals. These spirals are called **gyres**. Five major gyres circle the oceans: the North Atlantic, South Atlantic, South Pacific, North Pacific, and Indian Ocean. Each gyre has a western boundary current and an eastern boundary current.

The movement of water currents over Earth's surface acts as a "blanket." The constant movement of water keeps temperatures regulated on Earth. Warm water moves from the tropics toward the poles, and cold water moves from the poles toward the tropics. Water also evaporates from the surface of the oceans then falls back to Earth as rain and other forms of precipitation. Without wind and water currents, the areas near the tropics would be extremely hot and the areas near the poles would be extremely cold. Earth would not be a friendly place for the organisms that call it home.

Real-World Connection

There is no Coriolis effect at the Equator. Therefore, there are no ocean gyres at the Equator.

Climate Change

With changing climate, there is concern that the "global conveyor belt" of ocean currents may be disrupted. From increased rainfall to the melting of arctic ice, the sinking of cold water could be slowed, or even stopped. This could result in drastic temperature changes across the globe. This could, in turn, further slow the formation and sinking of cold water at the poles.

LESSON REVIEW

Complete the activities below to check your understanding of the lesson content. The Unit 4 Answer Key is on page 155.

Vocabulary

Write definitions in your own words for each of the key terms.

1. Coriolis effect ______________________________

____

2. current ______________________________

____

3. El Niño ______________________________

____

4. gyre ______________________________

____

Skills Practice

Answer the questions based on the content covered in the lesson.

5. What causes water currents to appear to move clockwise in the Northern Hemisphere and counterclockwise in the Southern Hemisphere?

   A. wind

   B. Earth's rotation

   C. uneven heating of Earth's surface

   D. melting of polar ice caps

6. Which process causes global wind systems to form?

   A. Earth's rotation

   B. uneven heating of Earth's surface

   C. movement of gyres in the oceans

   D. large-scale movement of polar ice

7. What is one way in which ocean currents help keep Earth's temperatures regulated?

   A. They prevent the ice caps from melting.

   B. They bring less rainfall to some parts of the planet.

   C. They move warm water from the Equator to the poles.

   D. They cause more water to be evaporated from the ocean surface.

INTERIOR STRUCTURE OF EARTH

Living on the surface of Earth, we may find it hard to imagine the layers below. However, we can sometimes see evidence of activity beneath the surface. Volcanic eruptions and earthquakes let us know that Earth is always moving and changing.

Key Terms

crust

inner core

mantle

outer core

Ring of Fire

subduction

tectonic plates

The Layers of Earth

Earth is a sphere that measures about 12,750 kilometers (km) in diameter. It is made up of three distinct layers: the core, the mantle, and the crust. We live on Earth's uppermost layer, called the **crust**. It is the thinnest of all Earth's layers, ranging from about 5 km to about 100 km deep. The crust includes the continents and the oceans.

Beneath the crust is the **mantle**, a layer much thicker than the crust, measuring about 2,900 km deep. The mantle is a hot, dense layer of semi-solid rock. The rock there is partially melted and flows under the crust.

The innermost layer of Earth is the core, and it is the thickest layer of Earth. It consists of two parts, the **outer core** and the **inner core**. The liquid outer core is 2,200 km thick and the solid inner core is 1,250 km thick. The core is also the hottest layer of Earth, because as depth increases, so do temperature and pressure.

Earth's Layers

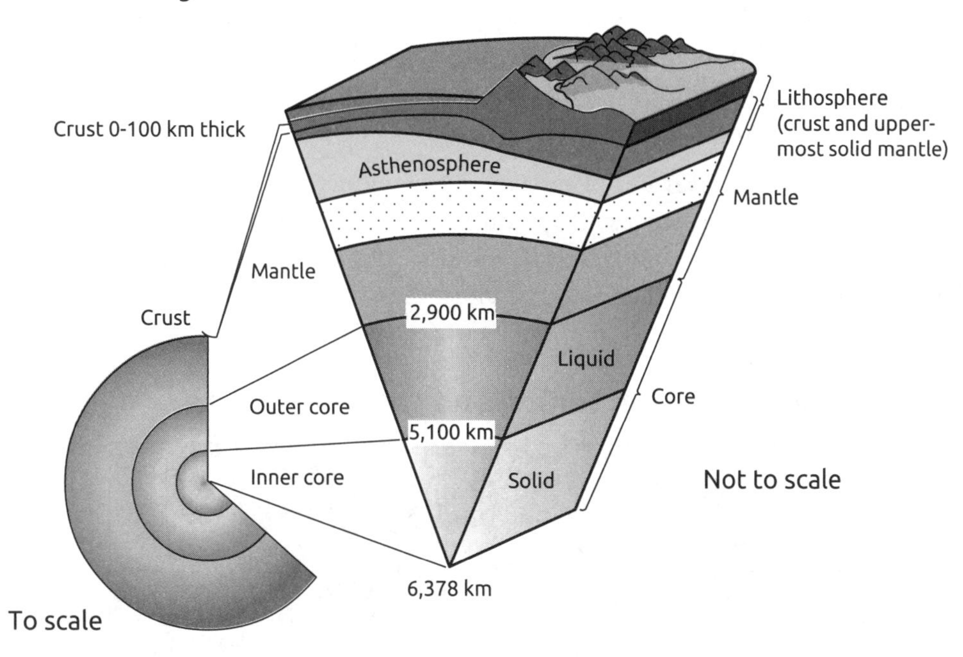

Source: USGS

Plate Tectonics

The upper part of the mantle and the crust form a layer of rock known as the lithosphere. The lithosphere is broken up into "plates" called **tectonic plates** that hold the continents and oceans. The plates are in motion, causing activity at their boundaries.

At divergent boundaries, two plates are moving apart and new crust is formed. At convergent boundaries, two plates collide. This can cause one plate to move under another, forcing old crust down into the mantle where it is melted and recycled. Two plates can also push against each other at a convergent boundary, creating a mountain range. At a transform plate boundary, two plates slide past each other. The crust could crack and break, but unlike at convergent and divergent boundaries, no crust is created or destroyed.

Types of Plate Boundaries

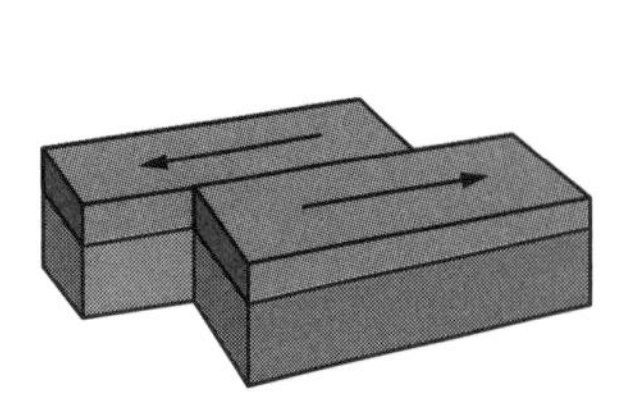

Transform

Divergent

Convergent

Earthquakes and Volcanoes

The activity of tectonic plates is difficult to detect without specialized equipment. But evidence of that activity is apparent all over the world. Both earthquakes and volcanoes are the result of plate tectonics. As the plates move, they grind against each other, causing the earth to shake. Sometimes, at a convergent plate boundary, one plate is forced under another plate. This is called **subduction** and occurs at a subduction zone. The plate that is shoved beneath melts and creates magma. The magma then comes up to the surface, resulting in an erupting volcano.

The rim of the Pacific Plate is an area of very active tectonic activity. About 90 percent of all earthquakes and 75 percent of all active volcanoes are located there. This area has been aptly named the **Ring of Fire**. More than 450 volcanoes dot the Ring of Fire, stretching from the southernmost part of South America, along the western coast of North America, across the Bering Strait, down Japan, and into New Zealand.

Real-World Connection

The Aleutian Islands in Alaska are a result of plate tectonics. As the Pacific Plate subducts, or is forced to slide under, the North American Plate, the islands continue to form. The Aleutian Islands contain 27 of the United States' 65 active volcanoes.

Active Volcanoes, Plate Tectonics, and the Ring of Fire

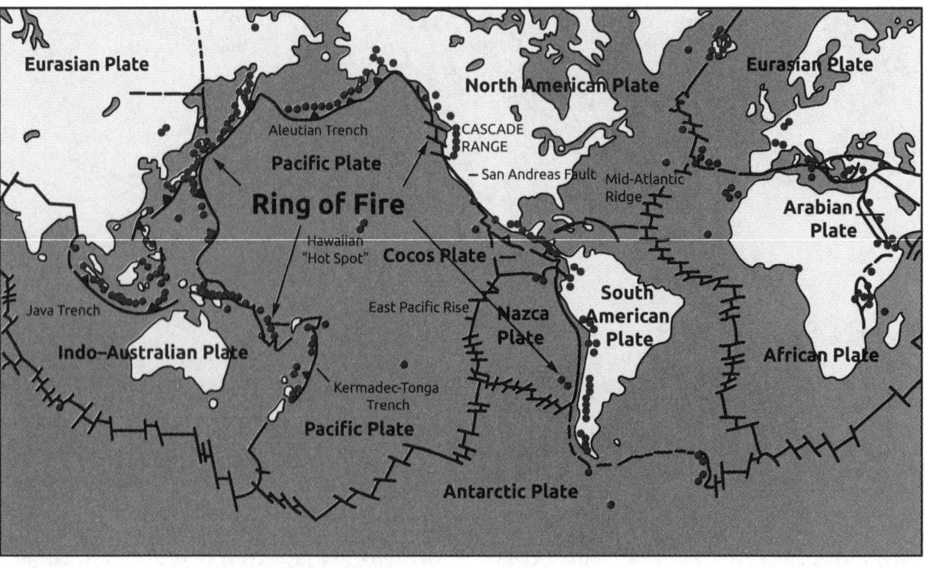

Source: USGS

LESSON REVIEW

Complete the activities below to check your understanding of the lesson content. The Unit 4 Answer Key is on page 155.

Skills Practice

Answer the questions based on the content covered in the lesson.

1. At a ________________________ boundary, crust is neither created nor destroyed.

2. The ________________________ is the thinnest of Earth's layers.

3. Which description most accurately describes the lithosphere?

   A. the crust only

   B. the mantle only

   C. the mantle and the lower part of the crust

   D. the crust and the uppermost part of the mantle

4. At which type of plate boundary are mountain ranges formed?

   A. convergent boundaries

   B. divergent boundaries

   C. mantle boundaries

   D. transform boundaries

5. Why is the Ring of Fire an area of many active volcanoes?

   A. Plate boundaries are constantly moving apart there, creating new crust.

   B. Two plates are grinding past each other, causing large earthquakes.

   C. The ocean is deepest there, allowing for the creation of large underwater mountain ranges.

   D. It is an area of subduction, causing the melting and movement of magma through the mantle and crust.

WEATHERING AND NATURAL HAZARDS

Key Terms

chemical weathering

earthquake

hurricane

physical weathering

weathering

As a mountain ages, like a person, its appearance may change. The mountain peak may become more rounded as rocks are worn away. The changes may be gradual and steady, or they may be abrupt and sudden. These variances do not occur just with mountains; landforms all over Earth are affected.

Weathering

Earth's surface and landforms change due to weathering. **Weathering** is the breaking down and wearing away of rocks. This is most often caused by water and wind. Weathering causes Earth's surface to change constantly.

Physical weathering causes rocks to break into smaller pieces. The composition of the rock is the same, but it changes physically into smaller pieces. Wind and water are major agents of physical weathering. The roots of plants can also cause physical weathering.

Wind, carrying dust and sand, blows against a rock or mountain, causing the rock to gradually break apart. Water then drips into a crack within the rock. The water freezes and expands, causing the crack to widen and eventually split the rock.

This diagram demonstrates how water can break apart a rock.

Physical Weathering by Water

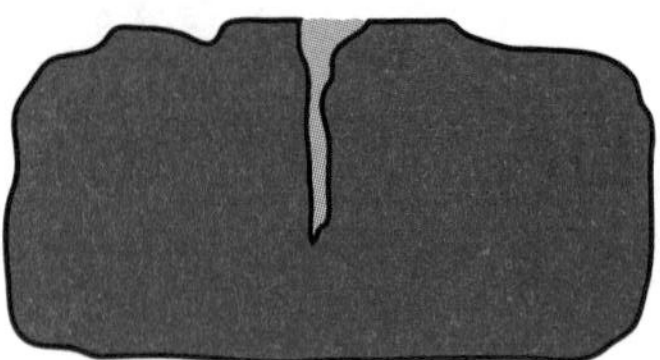

1. Water enters the cracks of a rock.

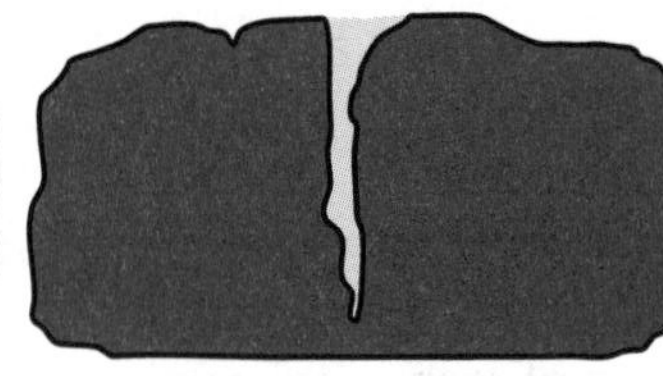

2. As temperatures drop, the water freezes within the crack. The ice expands, which begins wedging the rock apart.

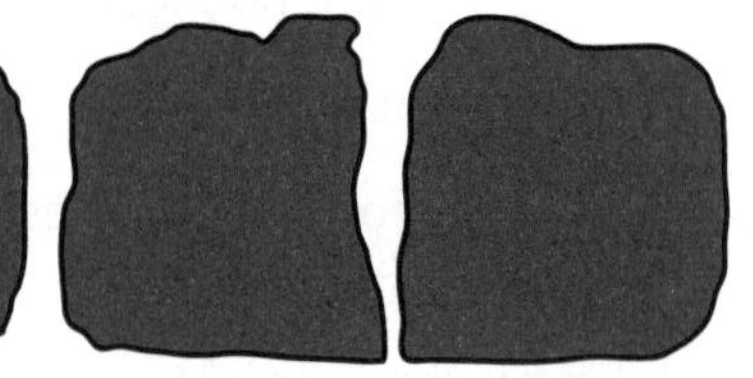

3. The water freezes and thaws repeatedly, and the rock splits.

Chemical weathering changes the composition of rocks by chemical reactions. Acid rain breaks down the rock, and the rock becomes weak and may dissolve. Water, carbon dioxide, and oxygen are agents of chemical weathering. Some types of stone, such as limestone, are much more prone to damage from chemical weathering.

Chemical weathering can dissolve rock and cause holes, as shown in the following image.

These holes are the result of chemical weathering.

Natural Hazards

Many examples of weathering are gradual and occur over time, while some changes occur more quickly. These rapid changes are often caused by natural hazards or disasters.

A **hurricane** is a large storm with winds greater than 74 mph. Because of the strong winds, a hurricane can change a landscape quickly, washing away rock and sediment and often causing trees to fall.

An **earthquake** is a sudden vibration in which plates slip past one another. It is often described as the ground shaking. This can cause great destruction and may lead to rockslides or landslides. Earthquakes, hurricanes, and other natural disasters cause very rapid weathering and changes to landforms.

Real-World Connection

Weathering is sometimes confused with erosion. Weathering breaks a rock down into smaller pieces, and erosion carries the pieces away.

LESSON REVIEW

Complete the activities below to check your understanding of the lesson content. The Unit 4 Answer Key is on page 155.

Vocabulary

Write definitions in your own words for each of the key terms.

1. chemical weathering ______________________________

____

2. earthquake ______________________________

____

3. hurricane ______________________________

____

4. physical weathering ______________________________

____

5. weathering ______________________________

____

Skills Practice

Answer the questions based on the content covered in the lesson.

6. Which would cause weathering to occur slowly?

   A. acid rain

   B. earthquake

   C. hurricane

   D. landslide

7. Which is a cause of physical weathering?

   A. acid rain

   B. carbon dioxide

   C. erosion

   D. water

8. Which answer correctly orders the following steps in explaining how ice breaks apart a rock?

   1. Water seeps into the crack of a rock.
   2. The water expands, cracking the rock.
   3. The crack expands many times, and the rock breaks apart.
   4. The water freezes.

   A. 2, 3, 1, 4

   B. 1, 4, 2, 3

   C. 3, 2, 4, 1

   D. 1, 2, 3, 4

AGE OF EARTH

Rocks provide some of the best evidence for calculating the age of Earth. They record the changes in the surface of Earth over the years. Scientists look among the rocks for evidence of chemical reactions, signs of erosion, and traces of animal and plant life to estimate Earth's age.

Key Terms

fossil

landform

radiometrics

The Age of the Earth

All over Earth, geologists have founds rocks that date back 3.5 billion years. Most of these rocks are the remains of lava flow from ancient volcanic eruptions. This means that Earth's age is even greater than 3.5 billion years. Radiometric dating techniques on certain crystals indicate an age of 4.3 billion years. Further evidence reveals an estimated age of 4.5 billion years, which marks the beginning of the formation of the solar system.

Radiometrics

Radiometrics are methods of using the decay rate of materials in rocks and soil to learn how long they have been there. Geologists have used radiometrics to estimate the age of Earth at 4.5 billion years. These techniques are also used to date fossils and determine the age of the organic matter that left the fossil imprint.

A common method for estimating the age of the remains of living things is carbon-14 dating. Carbon-14 is a weakly radioactive form of carbon that is found in small amounts in plants and animals. The amount of carbon-14 is constant in living organisms. When a living organism dies, the amount of carbon-14 inside it begins to decrease. Scientists can use this information to guess how long ago the organism died. Since carbon-14 decays very quickly, this method can only date objects that are many thousands, but not millions, of years old. Other substances, such as uranium-235, take much longer to decay and are used to date older materials like rocks and minerals.

Fossils

Fossils are the naturally preserved remains or traces of animal or plant life from the past. There are two types of fossils: body fossils and trace fossils. A body fossil is the actual body of an animal or plant that has been preserved in nature. A trace fossil is only the evidence of the presence of a living organism. An example of a trace fossil is a rock with the shape of a fish naturally imprinted on it.

Trace fossil of a fish

Landforms

Landforms are natural formations that occur on Earth. Landforms can be as large as continents, mountain ranges, plains, and coasts or as small as hills, valleys, and ponds. Landforms are shaped by natural elements, such as water and wind, or by the movement of the continents. For example, contact with water at coasts creates rocky or sandy landforms.

Scientists study landforms to gather information about the events that shaped the surface of Earth. An asteroid impact or the erosion of mountains can both destroy old landforms and create new ones. Landforms provide evidence of past events and are useful in determining the age of Earth.

Real-World Connection

Scientists have studied thousands of meteorites that have fallen to Earth to determine the age of the solar system. Using radiometrics, scientists estimate that the solar system was formed between 4.53 and 4.58 billion years ago. This result is consistent with other calculations that have determined the age of Earth at 4.5 billion years.

Unit 4 Lesson 5 LESSON REVIEW

Complete the activities below to check your understanding of the lesson content. *The Unit 4 Answer Key is on page 155.*

Vocabulary

Write definitions in your own words for each of the key terms.

1. fossil ______________________________

   ____

2. landforms ______________________________

   ____

3. radiometrics ______________________________

   ____

LESSON REVIEW

Skills Practice

Fill in the blanks.

4. The ______________________ dating method is used for relatively young, once-living matter.

5. A ______________________ fossil is the actual remains of an animal or plant that has been preserved in nature.

Answer the questions based on the content covered in the lesson.

6. What is the estimated age of Earth?

   A. 2.5 billion years

   B. 3.5 billion years

   C. 4.0 billion years

   D. 4.5 billion years

7. Which of the following can be used as evidence in determining the age of Earth?

   A. fossils

   B. rocks

   C. radiometrics

   D. meteorites

Provide examples for the following:

8. An example of a large landform ____

   ____

9. An example of a radiometric dating method ____

   ____

10. An example of a natural element that can cause erosion ____

   ____

STRUCTURES IN THE UNIVERSE

Key Terms

constellation

galaxy

solar system

star

The age of the universe is estimated at 14 billion years because the light that reaches us from the most distant objects appears to have been traveling for that long. Telescopes have helped us discover that the universe is expanding; all cosmic objects, such as stars and galaxies, are moving away from each other. Cosmologists believe that galaxies must have been much closer to each other in the past.

Stars

Stars are bright spheres of gas held together by their own gravity. They are born from clouds of matter scattered in the universe. Gravity pulls the gas particles that are in the clouds into a hot ball, where pressure builds. This pressure pushes outward and counteracts the force of gravity. The sun is the closest star to Earth, which is why its heat and light reach us. There are billions of stars in the universe, some similar in size to the sun and some much smaller or bigger.

Constellations

Constellations are patterns of stars in the sky that humans have observed and named. The Big Dipper is an example of a constellation. These stars appear close to each other from our point of view on Earth, but they are actually far away from each other in the universe. Different constellation patterns are identified based on the brightest stars observable from Earth.

The Big Dipper is one of the best-known constellations in the night sky.

Solar System

A **solar system** consists of a star and all the objects that orbit it. In our solar system, eight planets, their moons, asteroids, and other objects orbit the sun. Earth is the third-closest planet to the sun. The closest planet to the sun is Mercury, and the farthest one is Neptune.

Galaxies

A **galaxy** is a collection of dust, gas, and stars held together by gravity. A single galaxy can have at least one hundred billion stars in it. Our solar system resides inside the Milky Way galaxy, together with billions of other stars and solar systems. There are hundreds of billions of galaxies in the universe.

Our galaxy, the Milky Way, is a spiral galaxy like this one.

Real-World Connection

The universe is made up of galaxies, stars, planets, and other cosmic objects moving in space. Every object in the universe is constantly shifting form and starting a new phase. Some stars are born while other stars are dying. Galaxies move away from each other as new solar systems are forming within them. While it is difficult to determine the fate of the universe, scientists are certain that the structure of the universe and everything within it is always changing.

LESSON REVIEW

Complete the activities below to check your understanding of the lesson content. The Unit 4 Answer Key is on page 155.

Vocabulary

Write definitions in your own words for each of the key terms.

1. constellation ____

____

2. galaxy ____

____

3. solar system ____

____

4. star ____

____

Skills Practice

Answer the questions based on the content covered in the lesson.

5. Earth is contained within all of the following EXCEPT which one?

   A. the Milky Way

   B. the Big Dipper

   C. the solar system

   D. the universe

6. Which of the following is the biggest?

   A. a star

   B. a solar system

   C. a galaxy

   D. an asteroid

Fill in the blanks.

7. Since all cosmic objects are moving away from each other, cosmologists believe that the universe is ____________________.

8. In the formation of a star, ____________________ pulls together the particles into a hot ball of matter.

9. ____________________ is the farthest planet from the sun in the solar system.

STRUCTURES IN OUR SOLAR SYSTEM

Celestial bodies such as planets, moons, and suns are in constant motion. The relative positions of the sun and the moon can create events known to us as tides, seasons, and eclipses. We can predict when these events will occur because we know how these bodies move.

Key Terms

lunar eclipse

moon

planet

solar eclipse

tide

Planets

Planets are large, spherical bodies of matter that orbit a star. In our solar system, there are eight planets. Listed in order of closest to farthest from the sun, they are Mercury, Venus, Earth, Mars, Jupiter, Saturn, Uranus, and Neptune. While the inner four planets are made of rock, metal, and other solid matter, the outer four are made of gas. All the planets in our solar system orbit the sun. Earth is the only planet in our solar system that is known to have life.

Planets in the Solar System

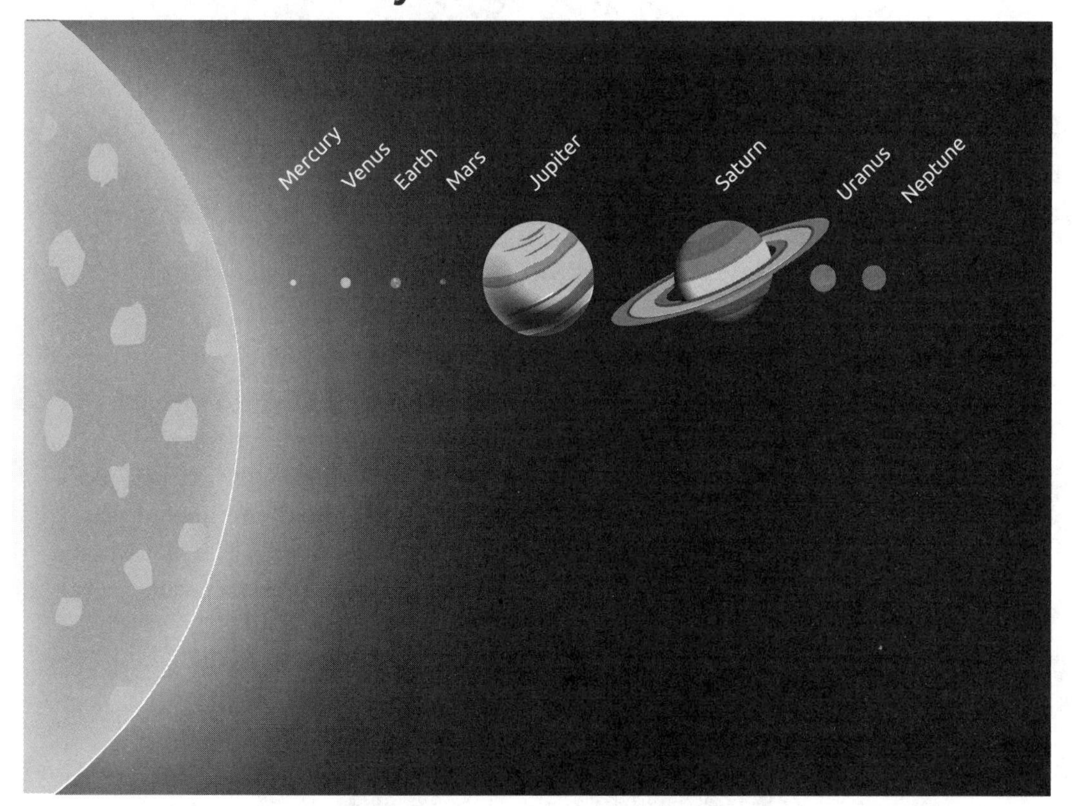

Source: NASA

Real-World Connection

Saturn, the sixth planet from the sun, is known for its distinct rings. From Galileo to the present, scientists have spent many hours observing and studying the rings of Saturn. Today, we know that the rings of Saturn are made of particles, some as small as a grain of sand and some as large as mountains, all of which orbit Saturn like tiny moons. Astronomers continue to study the rings of Saturn to understand their structure. Jupiter, Uranus, and Neptune also have rings, but theirs are much less prominent than Saturn's.

Moons

Moons are solid, spherical, and generally large objects that orbit the planets. Most large planets in our solar system have a few moons. Currently, there are more than 140 moons orbiting the eight planets in our solar system. No moons orbit Mercury or Venus, while Earth has only one moon. The four large gaseous planets capture many moons because of their large gravitational fields. Some of these moons may have frozen water or atmospheres on them.

Eclipses

An eclipse occurs when either the moon or Earth blocks the light or the view of the sun. During a **lunar eclipse**, Earth moves between the sun and the moon and casts a shadow on the moon. During a **solar eclipse**, the moon moves between Earth and the sun, where it blocks the view of the sun from Earth. A total eclipse occurs when the light or the view is completely blocked.

Lunar Eclipse

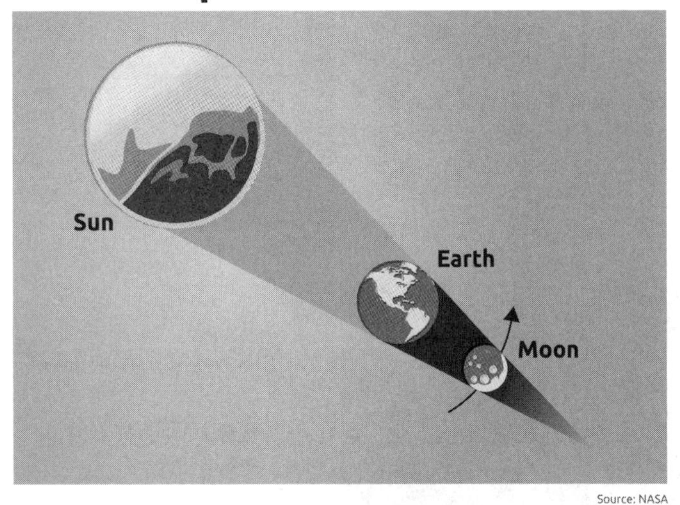

Source: NASA

Solar Eclipse

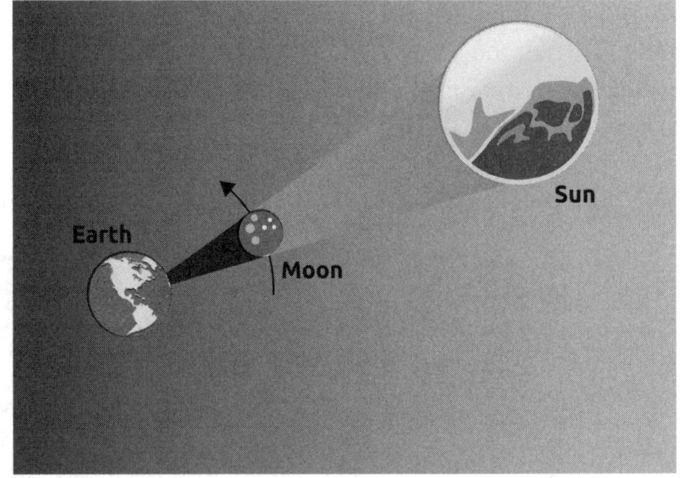

Source: NASA

Tides

Tides are the periodic rising and falling of water levels on Earth. This movement is mostly the result of the gravitational pull of the moon. Lunar tides occur when the moon's gravity pulls Earth's water toward it, creating a bulge in the section of the ocean directly underneath the moon. The effect that we see is certain coastal waters coming farther inland when the moon is overhead, then falling back when the moon is elsewhere in its orbit. Animals and plants that live in or near water have adjusted their behavior according to the tidal movement of the water.

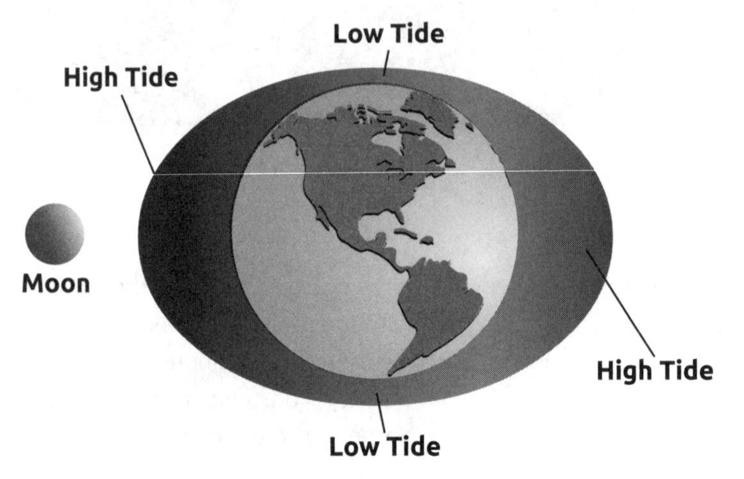

Tides are the result of gravitational pull of the moon.

LESSON REVIEW

Complete the activities below to check your understanding of the lesson content. The Unit 4 Answer Key is on page 157.

Vocabulary

Write definitions in your own words for each of the key terms.

1. lunar eclipse ______________________________

   ____

2. moon ______________________________

   ____

3. planet ______________________________

   ____

4. solar eclipse ______________________________

   ____

5. tide ______________________________

   ____

Skills Practice

Answer the questions based on the content covered in the lesson.

6. How many planets are in our solar system?

   A. four

   B. eight

   C. nine

   D. twelve

7. Which of the following planets has only one moon?

   A. Venus

   B. Mercury

   C. Earth

   D. Jupiter

8. Which of the following planets is made of gas?

   A. Saturn

   B. Mars

   C. Venus

   D. Mercury

UNIT PRACTICE TEST

Answer the questions based on the content covered in this unit. The Unit 4 Answer Key is on page 157.

1. A scientist is researching ways to reduce the amount of invasive algae in a nearby lake. What is one way the scientist could do this?

   A. Increase the number of trees near the lake, thereby decreasing the amount of carbon dioxide being released into the atmosphere near the water.

   B. Limit the amount of fertilizer used by farmers and local communities, thereby decreasing the amount of nitrogen in the water.

   C. Impose a restriction on the use of fossil fuels, thereby decreasing the amount of carbon in the carbon cycle.

   D. Add more predators of the algae into the lake, thereby decreasing the carbon dioxide and nitrogen produced by the algae.

2. Trees and other plants are being burned down in a forest ecosystem. What are two possible results of this deforestation in terms of the carbon cycle? Write your answer below.

   ____

   ____

   ____

   ____

   ____

3. Which statement describes how global winds form?

   A. Warm air rises at the Equator and moves toward the poles. Cooler air at the poles sinks and moves back toward the Equator.

   B. Warm air sinks at the Equator and moves toward the poles. Cooler air at the poles rises and moves back toward the Equator.

   C. Cool air rises at the poles and then sinks toward the Equator, where the warm air pushes up on the cooler air.

   D. Cool air sinks at the poles and then rises toward the Equator, where the warm air pushes down on the cooler air.

4. Why do global wind patterns curve as shown in the picture?

   The Coriolis Effect

   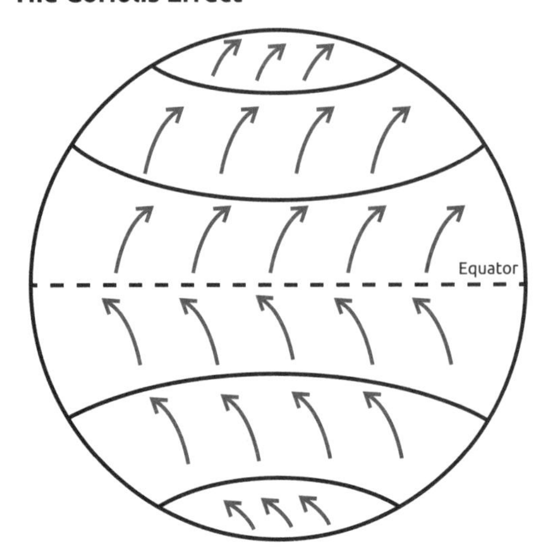

   A. As the air is moving in a cycle from the Equator to the poles, the earth is also rotating.

   B. The sun heats the earth unevenly, with the Equator receiving more direct sunlight than the poles.

   C. The gravitational force from the sun and the moon combine to pull the winds to the right or left.

   D. As the water is moving beneath the air, the weight of the water moves the air in spiral patterns.

5. It is often said that climate change could trigger a positive feedback loop. This means that one event can trigger another event, which would trigger more of the first event, continuing in a never-ending loop. How can the results of climate change in relation to ocean currents trigger a positive feedback loop? Write your answer below.

   ____

   ____

   ____

   ____

   ____

6. Why is rock in the mantle semi-solid?
   A. Volcanoes deep in the earth melt the rock.
   B. Old crust at subduction zones is pushed down.
   C. Temperature and pressure increase with depth.
   D. Earthquakes cause the heat in surrounding areas to rise.

7. A geologist is studying an area with frequent earthquakes. She observes that the land appears to be shifted, as shown in the image below.

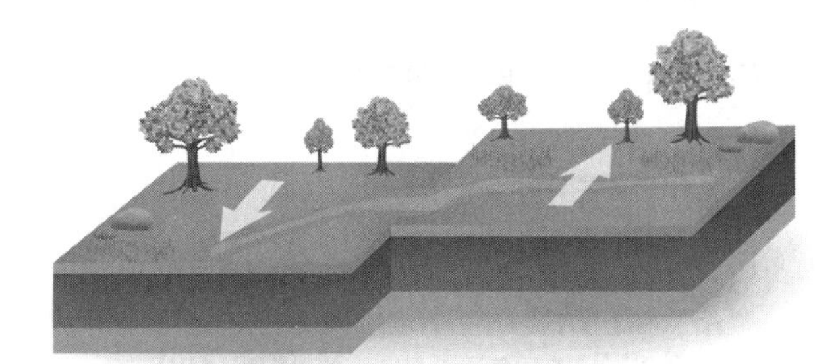

   What type of plate boundary is shown in the image?
   A. convergent
   B. divergent
   C. subduction
   D. transform

8. Explain how the carbon-14 radiometrics dating method is applied. Specify the type of material on which this method is used. Write your answer below.

____

____

____

____

____

9. Match the term to its correct example.

Term	Example
physical weathering	earthquake
chemical weathering	ice wedging
natural hazard	acid rain

10. Circle all the objects that are in our solar system:
    Mercury
    the Milky Way
    the moon
    the Big Dipper
    Neptune

11. What is the estimated age of the universe?
    A. 4.5 billion years
    B. 5 billion years
    C. 10 billion years
    D. 14 billion years

12. All the following planets are made of gas EXCEPT which one?
    A. Jupiter
    B. Saturn
    C. Mars
    D. Neptune

For questions 13–14, circle the correct option to complete each statement.

13. Both Mercury and Venus have [zero, two] moons.

14. During a [solar, lunar] eclipse, the moon moves between Earth and the sun and blocks the view of the sun from Earth.

POSTTEST

Answer these questions based on the content in this book.

Read the following passage. Then, answer questions 1–4.

A student has three pots. In each pot, he plants five lima bean seeds. Each pot contains a different type of soil: sandy, clay, and loamy. Each plant receives an equal amount of sunlight and water. The height of the plants is recorded in the following table.

Height of Plants (cm)			
Type of soil	**Day 5**	**Day 10**	**Day 20**
Sandy	2	6	9
Clay	1	4	6
Loamy	5	10	16

1. Which is a control variable in the investigation?
   - **A.** the kind of seed used
   - **B.** the heights of the bean plants
   - **C.** the type of soil in which the beans are planted
   - **D.** the amount of sunlight the plants receive

2. Which of the following could be a hypothesis for this investigation?
   - **A.** The plants in clay soil did not grow well.
   - **B.** If loamy soil is used, then bean plants will grow the tallest.
   - **C.** More pots for each soil type should have been used.
   - **D.** Lima bean seeds were not the best type of seed for this investigation.

3. The seeds came to the researcher already planted in the various types of soil and labeled in containers A, B, and C. The researcher did not know which type of soil was which. What was the purpose in setting up the experiment in this manner?
   - **A.** to reduce bias
   - **B.** to get accurate results
   - **C.** to prove the hypothesis
   - **D.** to make the experiment proceed faster

4. What was a weakness in this investigation?
   - **A.** One trial was conducted.
   - **B.** One type of seed was used.
   - **C.** There was no independent variable.
   - **D.** There was no way to repeat the results.

5. What is the purpose of a double blind study?
   - **A.** to reduce bias
   - **B.** to eliminate errors
   - **C.** to prove a hypothesis
   - **D.** to control the variables

Use the following table to answer question 6.

Height of Tomato Plant Grown in Soils	
Soil Sample	**Height (cm)**
1	6.5 cm
2	7.6 cm
3	8.2 cm
4	9.8 cm

6. What conclusion can be drawn from the data in the table?
   - **A.** Soil 1 cannot support tomato plant growth.
   - **B.** Soil 2 will hold the most water for tomato plants.
   - **C.** Soil 3 needed more sunlight and water for tomato plants to grow.
   - **D.** Soil 4 is the best for growing tomato plants.

Use the following graph to answer questions 7–10.

Tornadoes by Month: 1995–2013

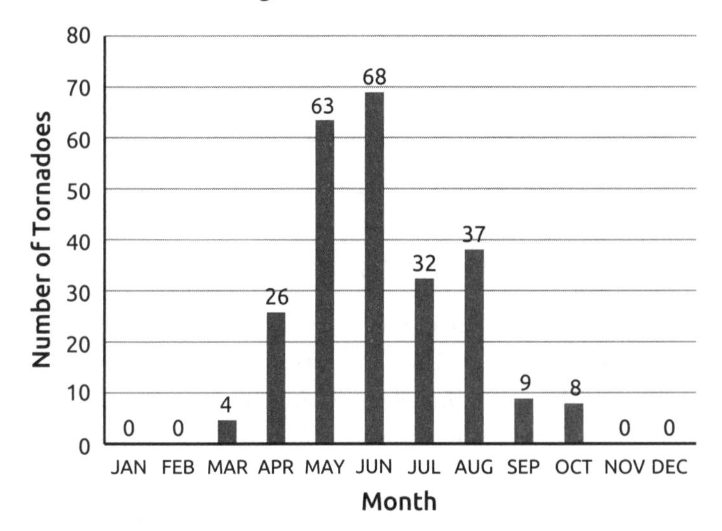

7. Which month had the fewest number of tornadoes?

   A. March
   B. June
   C. October
   D. December

8. Which statement about the number of tornadoes is true?

   A. Fewer tornadoes hit in June than in May.
   B. More tornadoes hit in July than in April.
   C. Fewer tornadoes came in March than in January.
   D. An equal number of tornadoes came in September and October.

9. What is the label on the *y*-axis of this graph?

   A. January–December
   B. Month
   C. Number of Tornadoes
   D. Tornadoes by Month

10. Which can be concluded by the results in the graph?

    A. The number of tornadoes increases during the winter.
    B. The number of tornadoes decreases during the summer.
    C. Fewer tornadoes occur in the winter than in the summer.
    D. More tornadoes occur in the spring months than in other months.

11. Which statement is an example of an inference?

    A. The patient's heart rate increased during exercise.
    B. The mouse ran on the wheel because it was bored.
    C. The tsunami occurred eight minutes after the earthquake.
    D. The ramp reduced the amount of work done on the box.

12. Which do plants NOT require in order to make their own food?

    A. oxygen
    B. water
    C. sunlight
    D. carbon dioxide

13. Which is produced during photosynthesis and used by animals and plants for respiration?

    A. water
    B. oxygen
    C. energy
    D. carbon dioxide

14. Yeast uses the sugar in grapes to produce alcohol through which process?

    A. germination
    B. fermentation
    C. photosynthesis
    D. respiration

15. A remora fish has attached itself to a leopard shark and is using the shark's energy to travel. The leopard shark is neither harmed nor benefitted by the remora fish. This is an example of

    ____

Use the following graph to answer questions 16–19.

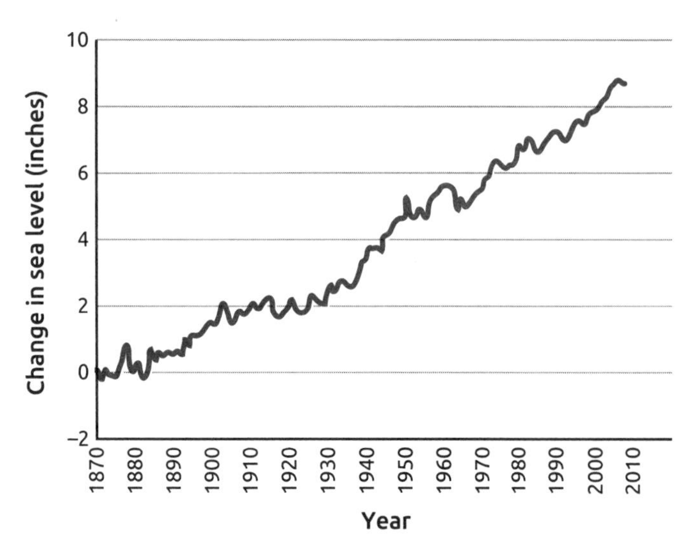

In this graph, the shaded line shows the likely change of sea level, which depends on the number of measurements and the methods used at different times.

Source: EPA's Climate Change Indicators (2012).

16. What can be concluded from the graph?

    A. Sea level has increased greatly since 1930.

    B. Sea level did not change between 1870 and 1920.

    C. The change in sea level has remained constant since 1890.

    D. The greatest change in sea level occurred between 1950 and 1960.

17. Describe the relationship between sea level change and time (the years between 1870 and 2010).

    ____

    ____

    ____

18. What prediction can be made based on the information in the graph?

    A. Sea level will decrease in the future.

    B. Sea level will become steady in the future.

    C. Sea level cannot be predicted for the future.

    D. Sea level will continue to increase in the future.

19. Much like using many subjects during a medical study, scientists use measurements from many different places on Earth when determining sea level. Explain why scientists follow this procedure.

    ____

    ____

    ____

Read the following passage. Then, answer questions 20 and 21.

A horse and a donkey are two different species, but they are able to produce offspring. A male donkey and a female horse can produce a mule, but the offspring will be unfertile; it cannot reproduce.

20. A horse has 64 chromosomes, and a donkey has 62 chromosomes. How many chromosomes does a mule have? ______________

21. Which statement best explains why the mule cannot reproduce?

    A. The mule does not have enough DNA to control the formation of gametes.

    B. So few mules are produced that reproductive isolation prevents them from mating.

    C. The mule is not fully developed; so it does not live long enough to reach reproductive age.

    D. During meiosis, differences in the chromosomes will prevent them from aligning, so the cells cannot divide.

22. Which statement best describes protein synthesis?

    A. The code of DNA is transcribed to form RNA, which carries the code to the ribosomes for the making of a protein.

    B. The code of RNA is translated to form DNA, which carries the code to the ribosomes for the making of a protein.

    C. The code of DNA is translated to form RNA, which carries the code to the ribosomes for the making of a protein.

    D. The code of RNA is transcribed to form DNA, which carries the code to the ribosomes for the making of a protein.

Read the following passage. Then, answer questions 23–24.

Lionfish are native to the Indian and Pacific Oceans. They are currently considered an invasive species in the Atlantic Ocean off the southeast coast of the United States, as well as in the Gulf of Mexico.

23. Which of the following BEST describes how the lionfish appeared in the Atlantic Ocean?

    A. They evolved there.

    B. They migrated there.

    C. They were carried there by currents.

    D. They were brought there by humans.

24. Name two possible consequences of lionfish living in the Atlantic Ocean and Gulf of Mexico.

    ____

    ____

    ____

25. When a person's arm is injured and becomes inflamed, white blood cells enter the tissue around the injury. They engulf any microbes and dirt that may have entered the body. This response is controlled by which systems working together?

    A. circulatory and immune

    B. endocrine and immune

    C. digestive and circulatory

    D. digestive and endocrine

26. Why is it important to receive vaccinations?

    ____

    ____

27. Allergies, such as those caused by pollen, occur when a specific type of immune cell recognizes that pollen is as a(n) ____________________ for which to create antibodies.

Read the following passage. Then, answer questions 28–29.

A man is running a mile-long race. After running for a few minutes, he starts to get tired and begins to breathe more heavily. He continues and finishes the race.

28. Which best explains why the person is breathing more heavily while running?

    A. His muscles need more energy, and oxygen is required to get that energy.

    B. His muscles need more sugar, and inhaling moves more sugar to the muscles.

    C. His muscles need more sugar, and exhaling moves more sugar to the muscles.

    D. His muscles need more energy, and carbon dioxide is required to get that energy.

29. Which statement best explains how body systems work together in the previous scenario?

    A. The digestive system takes in more oxygen, which is taken to the muscles by the excretory system.

    B. The circulatory system takes in more carbon dioxide, which is taken to the muscles by the respiratory system.

    C. The respiratory system takes in more oxygen, which is transported to the muscles by the circulatory system.

    D. The excretory system takes in more carbon dioxide, which is transported to the muscles by the digestive system.

30. Which system maintains a proper balance of substances in the body by releasing hormones?

    A. circulatory

    B. endocrine

    C. excretory

    D. nervous

POSTTEST

Read the following passage. Then, answer questions 31–33.

A student has crossed two types of tomato plants. One plant grows tomatoes that have smooth skin (S), and the other type of plant grows tomatoes with fuzzy skin (s). Smooth skin is dominant over fuzzy skin. The student uses a Punnett square as shown to determine what the offspring of these two parents would be.

	S	s
s	Ss	ss
s	Ss	ss

31. What percentage of offspring will have fuzzy skin?

    A. 25
    B. 50
    C. 75
    D. 100

32. What is the phenotype of a plant with the genotype Ss? ____________________

33. What are the alleles of a fuzzy-skinned tomato plant?

    ____

34. The reactions that occur in the making of new muscle are ____________________, so they ____________________ energy.

    A. catabolic, release
    B. catabolic, absorb
    C. anabolic, release
    D. anabolic, absorb

35. The average adult consumes 2,000 calories a day. Olympic gold medalist swimmer Michael Phelps consumes 12,000 calories a day. Why does he need to consume so many calories?

    ____

36. The lamprey eel attaches to a fish and sucks its blood. What type of behavior is this?

    A. parasitic
    B. predatory
    C. mutualistic
    D. communalistic

Use the following diagram to answer questions 37–39.

The following energy pyramid is representative of one that exists in Yellowstone National Park.

Partial Energy Pyramid in Yellowstone National Park

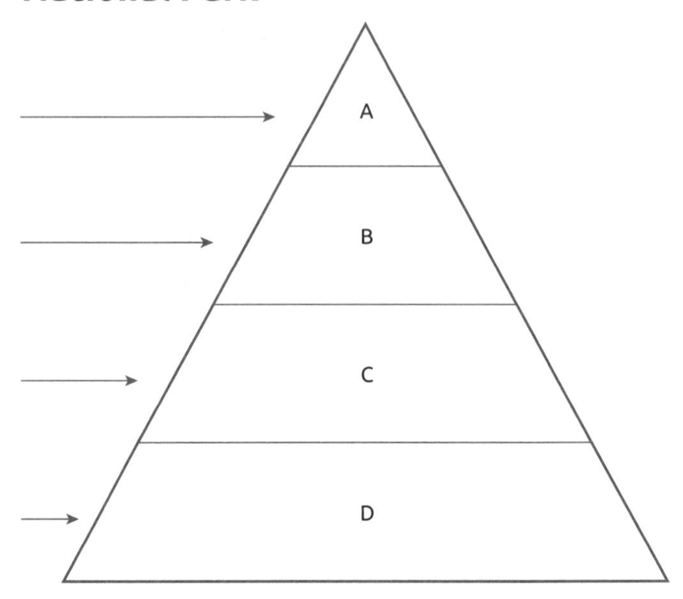

37. Match each organism to its location (A, B, C, D) in the energy pyramid.

    aspen trees ____________

    rabbits ____________

    wolves ____________

    deer ____________

38. In 1926, wolves were eliminated from Yellowstone Park. Describe what most likely happened to the populations of rabbits, deer, and aspen trees. Explain your reasoning.

    ____

    ____

    ____

39. Wolves eat small animals and rodents; similar to the food the birds of prey eat. If the population of wolves rises, the population of the birds of prey would ____________________.

Read the following passage. Then, answer questions 40–43.

Richard and Joan work for a package delivery company. They are delivering packages to a store that has both stairs and a ramp at the entrance. Richard's package has a mass of 20 kg, and Joan's package has a mass of 25 kg.

Potential energy is calculated using the equation $PE = mgh$, where g is the constant 9.8 m/s^2. Kinetic energy is calculated using the equation $KE = 0.5mv^2$. Acceleration is calculated using the equation $a = (v_f - v_i)/t$.

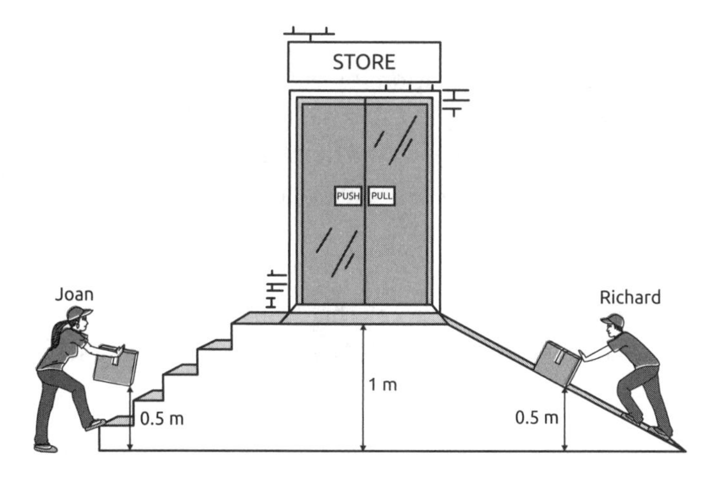

40. The potential energy of Joan's package is (more than, less than, equal to) the potential energy of Richard's package.

41. When Richard reaches the top of the ramp, he lets go of the package, and it begins to slide down the ramp. As the height of the package decreases, the speed at which it slides increases. What happens to the total energy of the package?

    A. The total energy of the package increases as it slides down the ramp.

    B. The total energy of the package decreases as it slides down the ramp.

    C. The kinetic energy of the package increases as it slides down the ramp, but the total energy remains the same.

    D. The kinetic energy of the package decreases as it slides down the ramp, but the total energy remains the same.

42. The starting velocity was 0 m/s, and the velocity after 2 seconds was 0.8 m/s. Richard's sliding package has an acceleration down the ramp of ____________________ m/s^2.

43. The ramp is an inclined plane, which is a simple machine. Name two other simple machines.

    ____

Read the following passage. Then, answer questions 44–46.

Sarah and Ruth are having a competition to see who is stronger. They each pulled a 40 kg cart at a constant force, which they measured with force gauges. Sarah pulled the cart with a force of 60N for 10 meters, while Ruth pulled the cart with a force of 20N for 35 meters.

$F = ma$ (mass times acceleration)

44. Acceleration can be calculated with the equation $a = F/m$. The acceleration of Sarah's cart is (more than, less than, equal to) the acceleration of Ruth's cart.

45. Work is calculated by the equation $W = Fd$ (force times distance). An onlooker mentioned that Sarah used more force but that Ruth did more work. Explain why the onlooker is right.

46. Sarah removed some weight from the cart, which gave it a mass of 30 kg. She managed an acceleration of 3.0m/s^2. How much force was she using to pull the cart?

    A. 10N

    B. 27N

    C. 60N

    D. 90N

47. ____________________, such as food and shelter, can have an effect on the carrying capacity of a population.

48. Evidence supports the idea that birds evolved from dinosaurs. This means that all birds have a common ____________________ that probably lived in the Cretaceous era.

49. Give two examples of renewable energy sources and two examples of non-renewable energy sources.

    Renewable: ______________________________

    Non-renewable: ______________________________

50. One advantage of (solar, nuclear) power is that it is not affected by the weather.

Draw a line from each object to the correct blank.

51. Match the type of electromagnetic radiation with its description.

ultraviolet	used in hospitals to see bones and internal body structures
infrared	component in sunlight that causes sunburns
X-ray	given off by warm objects, such as heat lamps used to keep food warm in restaurants

52. Visible light has a longer wavelength than X-rays. Longer wavelength radiation has (lower, higher) frequency and (lower, higher) energy than shorter wavelength radiation.

Read the following passage. Then, answer questions 53–57.

At the gym, a trainer keeps hand warmers for sore muscles and instant cold packs for swelling handy for her clients. The hand warmers use a reaction between iron powder (Fe) and oxygen gas (O_2) to produce iron oxide (Fe_2O_3). The cold packs work by dissolving a salt, such as ammonium nitrate (NH_4NO_3), in water (H_2O). The amounts of ammonium nitrate and water can be varied, depending on the desired temperature of the cold pack.

53. The reaction that produces iron oxide also produces heat, so it is (endothermic, exothermic), and dissolving ammonium nitrate in water absorbs heat, so it is (endothermic, exothermic).

54. Ammonium nitrate and water each have formulas that have more than one type of atom, so they are (compounds, elements). In the cold pack, they are combined physically, not chemically, so they form a (pure substance, mixture).

55. Heat travels from the hand warmer into the muscle it is touching by (conduction, convection, radiation) and is carried through the water in the cold pack by (conduction, convection, radiation).

56. The balanced equation for the reaction of iron powder and oxygen gas is $4Fe(s) + 3O_2(g) \rightarrow 2Fe_2O_3(s)$. This means the number of atoms of iron (Fe) will (decrease, stay the same, increase) during the reaction.

57. Explain how you can tell that the hand warmers use a chemical change, while the instant ice packs use a physical change.

    ____

    ____

    ____

58. Which pH indicates the most acidic solution?

    A. 3 **C.** 11

    B. 7 **D.** 14

59. A chemistry student mixed potassium chloride into water at 25°C. He kept adding more potassium chloride until no more would dissolve. The solution, at that point, is (unsaturated, saturated, supersaturated).

60. When carbon powder and oxygen gas are mixed, no chemical reaction occurs. Describe the difference between that mixture and the compound carbon dioxide (CO_2).

    ____

    ____

    ____

61. A scientist is studying a new compound, which has a melting point of –15°C and a boiling point of 128°C. At 0°C, the compound is a (solid, liquid, gas).

62. Sulfur has an atomic number of 16 and an average atomic mass of about 31 amu, while carbon has an atomic number of 6 and an average atomic mass of about 12 amu. This demonstrates the trend on the periodic table that average atomic mass (decreases, stays the same, increases) with increasing atomic number.

63. Which is NOT a way that carbon dioxide enters the atmosphere?

A. plant respiration

B. burning gasoline in a truck engine

C. photosynthesis

D. a forest fire

64. How do scientists estimate the age of the universe?

A. by using carbon-14 dating to estimate the age of asteroids that hit the Earth

B. by using radiometrics to estimate the age of meteorites that have fallen to Earth

C. by calculating the time it took for light to reach the Earth from the most distant objects in the universe

D. by measuring the rate at which stars die and new stars are born

65. Which of these is a trace fossil?

A. a dinosaur footprint

B. a petrified tree

C. a dinosaur skeleton

D. an insect trapped in amber

66. Why do the gaseous planets in our solar system have more moons than the solid planets?

A. The solid planets have magnetic fields.

B. The solid planets are closer together.

C. The gaseous planets are farther from the sun.

D. The gaseous planets have larger gravitational fields.

67. What is the primary cause of the daily tides in Earth's oceans?

A. the pull of the moon's gravity on the oceans, combined with Earth's rotation

B. the pull of the moon's gravity on the oceans, combined with the movement of the moon around Earth

C. the pull of the sun's gravity on the oceans, combined with Earth's rotation

D. the pull of the sun's gravity on the oceans, combined with the movement of Earth around the sun

68. At which type of plate boundary is new crust formed?

A. transform

B. convergent

C. tectonic

D. divergent

69. Which event is NOT caused by the movement of celestial bodies, such as planets, moons, and stars?

A. tides

B. earthquakes

C. eclipses

D. seasons

70. Earth's center is about how far from its surface?

A. 64 km

B. 640 km

C. 6,400 km

D. 64,000 km

71. Which of these is an example of erosion?
    - **A.** Acid rain reacts with a rock and gradually weakens it.
    - **B.** Waves beating on the seashore turn rocks into sand.
    - **C.** Water freezing and thawing creates a large crack in a rock.
    - **D.** A flood washes away sand from a river bank.

72. The following diagram shows the position of the moon in relation to Earth. The numbers refer to different places on Earth's surface.

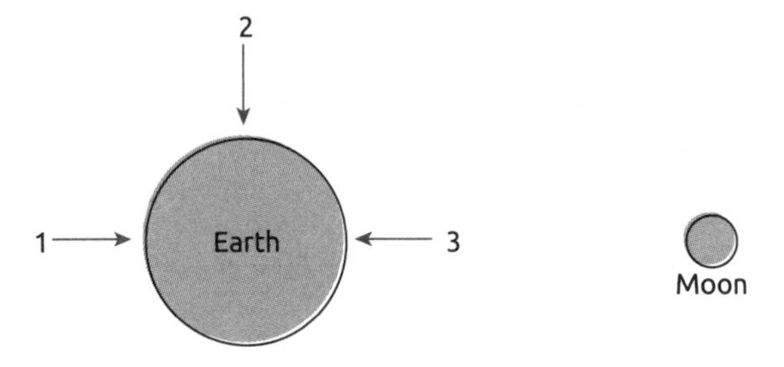

Which point or points would experience low tide?

- **A.** 1
- **B.** 2
- **C.** 3
- **D.** 1 and 3

73. Which of these is NOT a process by which landforms can be created?
    - **A.** asteroid impact
    - **B.** erosion by water
    - **C.** wind
    - **D.** mining

Choose the correct answer for questions 74–77 from the choices provided.

warmer	global winds
cooler	Earth's rotation

74. Global winds form because the sun heats the Earth unevenly. Air near the equator is warmer, and air near poles is ______________________.

75. As air moves from the equator toward the South Pole, the Coriolis effect causes winds to seem to move counterclockwise. The Coriolis effect is caused by ______________________.

76. The global ocean currents called gyres are caused by ______________________.

77. El Niño is a global climate condition that occurs when the tropical Pacific Ocean is ______________________ than usual.

78. About how old is Earth?
    - **A.** 1,000,000 years
    - **B.** 450,000,000 years
    - **C.** 10,000,000,000 years
    - **D.** 4,500,000,000 years

79. Write the objects in the left column on the lines in the right column in order from largest to smallest.

Earth	**largest**
Jupiter	______________
the Milky Way	______________
the moon	______________
the solar system	the sun
the universe	______________
	______________
	______________
	smallest

80. The photograph below shows an example of what type of weathering?

____

81. Describe how a divergent boundary between two tectonic plates is different from a transform boundary in terms of plate movement.

____

____

____

82. Describe how physical weathering and chemical weathering are different.

____

____

____

In the following questions, you will describe how nitrogen moves through different parts of an ecosystem in a cycle, based on the diagram of the nitrogen cycle shown. Use this diagram to answer questions 83–86.

The Nitrogen Cycle

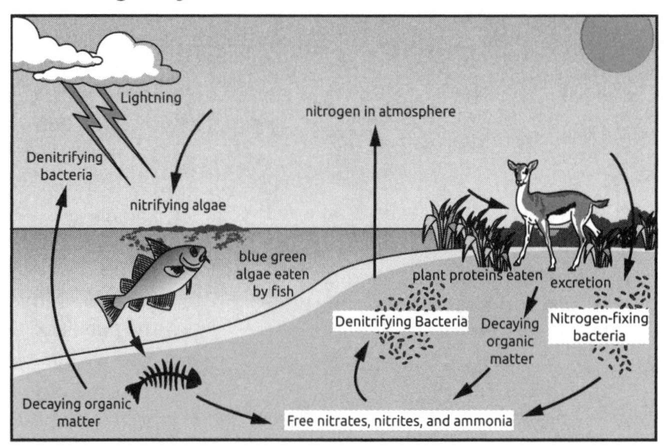

83. Describe one way nitrogen can move from the air into the soil.

____

____

84. Describe one way that nitrogen can move from the forest soil into a deer's body.

____

____

85. Describe one way that nitrogen can move from a deer's body into the forest soil.

____

____

86. Describe one way that nitrogen can return from the forest soil to the atmosphere.

____

____

POSTTEST ANSWER KEY

1. **D.** Control variables are the variables that are kept constant in an experiment, including the amount of sunlight and water plants receive.
2. **B.**
3. **A.** When the researcher does not know which soil is which, it eliminates any preconceptions or bias the researcher might have. This makes the results more accurate.
4. **A.** Investigations should have repeated trials to strengthen the results.
5. **A.**
6. **D.** Conclusions can only be drawn about which plants grew the tallest.
7. **D.** December had no tornadoes, while at least one tornado hit during the other listed months.
8. **B.**
9. **C.** The *y*-axis is along the vertical axis of the graph.
10. **C.**
11. **B.** An inference is a conclusion based on evidence.
12. **A.** Plants require oxygen for respiration, but not for photosynthesis. Plants need sunlight, carbon dioxide, and water for photosynthesis.
13. **B.**
14. **B.**
15. **commensalism** In commensalism, one species benefits, and the other is not affected. In the situation described, the remora fish is benefitting, while the shark is unaffected.
16. **A.** The graph shows a line with a generally positive slope, meaning an increasing trend.
17. **Sea level change and time are directly related; as the year increases, sea level height increases.**
18. **D.**
19. **The higher the number of results/measurements, the more accurate the average will be.** Having more data in an investigation/study leads to more accurate results.
20. **63** During meiosis, the chromosome number is cut in half, so the gametes of each parent have half the chromosome number. A horse will provide 32 chromosomes, and the donkey will provide 31 chromosomes, for a total of 63 chromosomes in the offspring.
21. **D.**
22. **A.** During the process of protein synthesis, the DNA is first transcribed to mRNA. The mRNA leaves the nucleus and goes to a ribosome. At the ribosome, translation occurs, and amino acids bond together in an order set by the mRNA. The amino acids will bond together to form a protein.
23. **D.** Because these fish are considered to be invasive, they would not have migrated to this area nor evolved here. Additionally, if they were carried by currents, they would have been in the area before, so they would not be considered invasive. Therefore, they were most likely brought in by humans.
24. **Sample answer: With no natural predators, they can overpopulate the region. This could lead to fewer fish for other predators to eat and also decrease the populations of what the lionfish eat.**
25. **A.** The white blood cells are carried by the circulatory system; because they fight infection in the body, they are also part of the immune system.
26. **Sample answer: Vaccinations allow your body to build up immunity to a pathogen through the creation of memory cells; these cells will be able to quickly identify the pathogen should it enter the body again.**
27. **antigen** Allergens have antigens that the body recognizes as foreign. The body will produce antibodies that are specific to the antigen. This starts an immune response.
28. **A.** When exercising, the muscles require a lot of energy. The process by which the cells of the body release the energy stored in food is respiration. Oxygen is required for respiration to occur.
29. **C.** The muscles require more energy. This energy is obtained through respiration. Respiration requires oxygen, which enters the body through the respiratory system. The oxygen is then transported through the body by the circulatory system.
30. **B.**
31. **B.** If a homozygous recessive plant is crossed with a heterozygous plant, half the offspring will produce smooth-skinned tomatoes, and half will produce fuzzy-skinned tomatoes.
32. **smooth-skinned tomatoes** The plants that produce smooth tomatoes could be SS or Ss since smooth skin is dominant.
33. **The alleles are ss.** The alleles are the versions of a trait that parents can pass on to offspring. A parent with fuzzy skin has both recessive alleles for the trait.
34. **D.** Anabolic reactions are those in which substances (such as muscles) are built up; these reactions absorb energy.
35. **Sample answer: He uses a lot of energy training, so he needs to take in more energy in the form of calories.**
36. **A.** When one animal uses another animal to the detriment of that animal, this is parasitic behavior. In this case, the lamprey eel is the parasite.
37. **aspen trees D., deer B., rabbits C., wolves A.**
38. **Without the wolves as apex predators, the coyote populations increased, as did the deer populations. Because there were more deer eating the plants, such as the aspen trees, their populations decreased.**
39. **decrease**
40. **more than** PE = mgh; the packages are at the same height, and the acceleration of gravity is the same, but Joan's package has more mass.
41. **C.** Total energy must remain the same, because of the law of conservation of energy. The package loses potential energy as the height decreases, so the kinetic energy must increase.
42. **0.4** $a = (v_f - v_i)/t = (0.8 - 0)/2 = 0.4$ m/s^2
43. **wedge, lever, wheel and axle, pulley, screw**
44. **more than** $F = ma$, so $a = F/m$. The mass is the same, but Sarah used more force, so the acceleration of her cart will be greater.
45. **Sarah's force of 60N is greater than Ruth's force of 20N, but Sarah did 60N x 10m = 600J of work, while Ruth did 20N x 35m = 700J of work.**
46. **D.** $F = ma$ = 30 kg x 3.0m/s^2 = 90N
47. **Limiting factors** Limiting factors are those factors that organisms need to survive, including food, water, and shelter.
48. **ancestor** A common ancestor is the most recent individual from which all organisms are descended.
49. **Sample answer: Renewable: solar, wind, hydroelectric; Non-renewable: coal, natural gas, nuclear**
50. **nuclear** Solar power depends on getting energy from the sun, which can be blocked by clouds and rain.
51. **ultraviolet: component in sunlight that causes sunburns; infrared: given off by warm objects, such as heat lamps used to keep food warm in restaurants; x-ray: used in hospitals to see bones and internal body structures**

52. lower; lower
53. **exothermic; endothermic** "Exo" means "out," and "thermic" relates to heat. Something that puts out heat to its surroundings will be exothermic.
54. **compounds; mixture** Both ammonium nitrate and water have formulas with more than one element in them, so they are compounds. Dissolving a compound does not involve making a new substance, so the result is a mixture.
55. **conduction, convection** Conduction involves heat traveling through materials that are touching each other, while convection involves heat traveling by liquid or gas carrying the heat.
56. **stay the same** According to the Law of Conservation of Mass, atoms can't be created or destroyed. There are 4 Fe atoms in 2 Fe_2O_3.
57. Sample answer: A new substance is produced in the hand warmers, one different from the reactants, but the chemical identities of the ammonium nitrate and water do not change, as they just form a mixture of the two compounds.
58. **A.** The pH scale runs from 0 to 14 and from acid to base, so the lower number will indicate the most acidic solution.
59. saturated
60. Sample answer: The compound carbon dioxide is made of molecules that have one atom of carbon and two atoms of oxygen; they formed the bonds during a chemical reaction, and the 1:2 ratio cannot change. Carbon dioxide gas has different properties from either carbon powder or oxygen gas. The mixture of carbon powder and oxygen gas is not chemically reacted and can be any ratio. The properties of the carbon powder and the oxygen gas do not change.
61. liquid
62. increases
63. C.
64. C.
65. A.
66. **D.** A planet gains moons when debris orbiting the planet comes together or when an object passes near enough to the planet to be captured by its gravitational field. Both of these events are more likely when a planet has more mass, which gives it a larger gravitational field.
67. A.
68. B.
69. B.
70. C.
71. **D.** Weathering can break rocks down physically or transform them chemically. Erosion occurs when the broken-down rocks are carried away.
72. B.
73. **D.** Landforms are created by natural processes. All of the answers except mining are natural processes that change the landscape.
74. cooler
75. Earth's rotation
76. global winds
77. warmer
78. D.
79. The complete order will be: the universe, the Milky Way, the solar system, the sun, Jupiter, Earth, the moon.
80. **physical** Plant roots can cause physical weathering, which breaks rocks into smaller pieces without changing their composition.
81. Sample answer: At a divergent boundary, two tectonic plates are moving away from each other. At a transform boundary, two plates are sliding past each other.
82. Sample answer: In physical weathering, the rocks are broken down, but their chemical composition remains the same. In chemical weathering, the rock is broken down and also changes composition.
83. Sample answer: Nitrogen-fixing bacteria in the soil turn nitrogen gas from the air into nitrates in the soil.
84. Sample answer: Plants take up nitrates from the soil and use them to make proteins. The deer eats the plants and incorporates the nitrogen into its own body.
85. Sample answer: After the deer dies, its body decays, returning nitrogen to the soil.
86. Sample answer: Denitrifying bacteria take up nitrates from the soil and convert them into nitrogen gas, which is released to the atmosphere.

After checking your Posttest answers using the Answer Key, use the chart below to find the questions you did not answer correctly. Then locate the pages in this book where you can review the content needed to answer those questions correctly.

Question	Where to Look for Help		
	Unit	Lesson	Page
1, 2, 19	1	1	13
3, 4, 5	1	2	15
6, 10, 11	1	4	19
7, 8, 9	1	3	17
12, 13, 14, 35	2	9	54
15, 36	2	12	66
16, 17, 18	1	5	21
20, 22	2	3	32
21	2	2	28
23, 24, 38, 39, 47	2	11	63
25, 28, 29, 30	2	7	47
26, 27	2	8	51
31, 32	2	4	36
33	2	5	39
34	2	1	25
37	2	10	58
40, 41	3	1	73
42, 44, 46	3	5	86
43, 45	3	7	93
48	2	6	42
49, 50	3	2	76
51, 52	3	4	83
53, 55	3	3	80
54	3	8	96
56, 60, 62	3	10	103
57, 61	3	9	100
58, 59	3	11	108
63, 83, 84, 85, 86	4	1	116
64, 79	4	6	132
65, 73, 78	4	5	129
66, 67, 72	4	7	135
68, 69, 70, 81	4	3	123
71, 80, 82	4	4	126
74, 75, 76, 77	4	2	120

ANSWER KEY

Unit 1: Science Reasoning

Lesson 1: Investigation Design

1. C.
2. A.
3. B.

Lesson 2: Evaluation of Investigations

1. anything that improperly influences the results or interpretation of an experiment
2. a type of experimental design meant to eliminate bias, wherein neither the researcher nor the patient knows whom is receiving treatment
3. a flaw or mistake made during a scientific investigation, possibly through inaccurate measuring tools or poor sampling techniques
4. D.
5. C.

Lesson 3: Comprehending Scientific Presentations

1. July 15
2. outside

Lesson 4: Reasoning from Data

1. They study their data and use evidence to make a statement about the investigation.
2. C.
3. Sample answer: He can conclude that as temperature increases, pressure increases. He can predict that at 85°C, the pressure will be 1.19 atm.

Lesson 5: Expressing Scientific Information

1. 45,000 feet
2. test flight 3
3. B.

Unit Practice Test

1. dependent
2. question/topic: Do pill bugs prefer dark or light locations?
   hypothesis: 75% of the pill bugs will stay in the dark.
   independent variable: the amount of light
   control variable: the time increments measured
3. Sample answer: The heat from the light might create a difference in temperature, which could attract or repel the pill bugs.
4. Sample answer: This is a very small group of subjects and should be repeated with a much larger group.
5. C.
6. B.
7. A.
8. The amount of each substance dissolved increases as temperature increases. However, more sugar can be dissolved than salt. Therefore, sugar is more soluble than salt.
9. D.
10. B.

Unit 2: Life Science

Lesson 1: Essential Functions and Components of Life

1. the basic unit of life
2. In a chemical reaction, the atoms in one or more substances rearrange and change how they are connected to each other to produce at least one new substance.
3. a large molecule that speeds up a chemical reaction within a living organism
4. the sum of all chemical reactions that take place within a living organism to maintain life
5. A, C, D.
6. C.
7. nucleus
8. product
9. cell membrane
10. catabolism

Lesson 2: Cell Theory and Growth

1. the process, from beginning to end, of the division of a cell into new identical cells
2. the thread-like structure that contains DNA in the nucleus of a cell
3. each functional part within a cell
4. Mitosis is a form of cell division by which a single cell divides into two identical daughter cells.
5. Meiosis is a form of cell division by which a single cell divides into four daughter cells. Each of the daughter cells resulting from meiosis carries only half of the chromosomes of the original cell.
6. cytoplasm
7. 46
8. mitosis
9. meiosis
10. skin
11. D.
12. First, it holds on to the genetic information of the cell that is stored inside the DNA. Second, it monitors the activities of the cell such as growth, metabolism, and reproduction.
13. A, C.
14. tissue, organ, system
15. Possible answers: circulatory, digestive, skeletal, immune, nervous, muscular
16. Possible answers: heart, brain, liver, skin, stomach
17. A, C.

Lesson 3: DNA and Chromosomes

1. a building block of protein; humans use 20 amino acids
2. a long strand of DNA containing many genes; humans have 23
3. three bases in a row on an RNA strand that code for an amino acid
4. a double-stranded molecule containing sequences of the bases A, T, G, and C
5. a segment of a DNA strand containing the instructions for a certain protein
6. a molecule used to build cells, send instructions, and help in cell functions
7. a single-stranded molecule similar to DNA, using U instead of T
8. copying a gene sequence from DNA to RNA
9. connecting amino acids to form proteins, following an RNA sequence
10. TTGACTAATGT
11. UUGACUAAUGU
12. C.

Lesson 4: Mechanics of Inheritance

1. the set of genes within DNA that are responsible for specific traits
2. the expression of a physical characteristic or trait set by the genes
3. a diagram used to predict the outcome of breeding in relation to the genotypes of specific traits
4. Tt
5. D.

Lesson 5: Changes in DNA

1. C.
2. D.
3. two or both

Lesson 6: Evolution

1. a trait that gives a species an improved function to allow it to survive

2. an organism to which all members of a species can trace their lineage
3. the process by which individuals with traits that are best suited to a species' environment survive to pass them on
4. the process by which one species becomes two or more new species
5. the different traits that exist within a species
6. Sample answer: Certain traits within a species allow more of the individuals with those traits to survive in their environment. Eventually, most of the individuals will have this trait, making it an adaptation.
7. C.
8. common ancestor
9. D.

Lesson 7: Body Systems and Homeostasis

1. digestive
2. respiratory
3. muscular
4. circulatory
5. homeostasis
6. D.
7. D.
8. A.
9. B.
10. B.
11. C.
12. B.
13. A.

Lesson 8: Disease and Pathogens

1. pathogen
2. toxin
3. virus
4. bacteria
5. infection
6. skin
7. immune
8. bacteria
9. D.
10. A.

Lesson 9: Nutrients and Energy

1. the building block of proteins in our bodies
2. a measurement of the potential of a food for producing energy in our body
3. the process of producing energy when there is not enough oxygen to break down sugar
4. a process in plants that uses sunlight to produce sugar
5. the process of consuming oxygen and releasing carbon dioxide
6. inorganic
7. fat-soluble
8. 11
9. C.
10. C.

Lesson 10: Flow of Energy and Matter in Ecosystems

1. a community of living organisms and nonliving things that work together to make a balanced system
2. a model of feeding connections among the species in an ecosystem
3. the feeding relationships between the organisms in an ecosystem
4. living: trees, frogs, owls nonliving: rocks, water, sunlight
5. Producers are organisms that capture the energy of the sun to make food. Examples: grass, trees, plants
6. An energy pyramid is a model of the flow of energy in an ecosystem. The shape of the pyramid displays how the amount of useful energy that enters each new trophic level decreases.
7. C.
8. Decomposers
9. Tertiary
10. food chain
11. grass
12. Sample answers: mouse, grasshopper, rabbit
13. Sample answers: hawk, toad, garter snake, mouse
14. Sample answers: hawk, mouse, hognose snake, sparrow
15. From bottom to top: wheat grains, sparrow, small snake, owl

Lesson 11: Carrying Capacity and Disruptions of Ecosystems

1. the maximum population an ecosystem can support without losing resources
2. a plant or animal that is not native to an area and has a negative effect on that area
3. a factor that limits the size of a population
4. C.
5. plant availability for food, size of snake population

Lesson 12: Symbiosis

1. B.
2. The mackerel population would increase because there would be fewer tuna feeding on them.
3. Sample answer: This is an example of commensalism. The monarch butterfly benefits from the chemical in the milkweed plant. The milkweed plant is not helped or harmed.
4. D.

Unit Practice Test

1. D.
2. A.
3. cell: red blood, white blood
tissue: bone, muscle
organ: heart, pancreas
system: respiratory, immune
4. meiosis
5. mitosis
6. mitosis
7. meiosis
8. codons
9. transcription
10. Proteins, amino acids
11. AAGGTGCAGCCGTAG
12. val-phe-ile
13. phenotype of offspring: yellow seed
genotype of offspring: YG
dominant gene: Y
recessive gene: G
14. The genotype of the other parent is Tt. Because one offspring has two recessive genes and one has two dominant genes, both parents must have one of each of these genes as well.
15. Sample answer: The epigenome sends signals that tell certain genes to turn on or off in a cell. If a gene for a certain function is not needed, then the epigenome acts like an on-off switch for that gene.
16. Sample answer: Alleles vary from person to person. Mutation causes different alleles.
17. D.
18. The green walking sticks and the striped walking sticks would not be able to produce offspring.
19. A.
20. B, D.
21. C.
22. D.
23. C.
24. B.
25. C.
26. A.

ANSWER KEY

Unit 3: Physical Science

Lesson 1: Types of Energy and Transformations

1. Possible answer: Mechanical energy is the total amount of kinetic and potential energy an object has.
2. potential
3. kinetic
4. potential
5. kinetic
6. D.
7. A.
8. D.

Lesson 2: Sources of Energy

1. Possible answers: Fossils fuels are readily available and relatively cheap. Power plants are not dependent on weather.
2. Possible answers: They cannot be used up. They do not pollute the air. They do not contribute to climate change.
3. A.
4. C.
5. nuclear
6. Photons
7. weather
8. turbine

Lesson 3: Flow of Energy

1. the movement of heat through a substance by direct contact
2. the motion of a gas or liquid whereby warmed air rises as it becomes less dense and sinks as it cools, creating a rotating current
3. heat energy
4. a measure of the kinetic energy of a substance
5. A.
6. C.
7. endothermic
8. exothermic
9. exothermic
10. kinetic

Lesson 4: Waves

1. the entire spectrum of radiation, ranging from radio waves to gamma rays
2. how often the particles vibrate as a wave passes through a medium
3. the release of energy as an electromagnetic wave
4. the distance between two successive points in a wave
5. B.
6. A.
7. ultraviolet

Lesson 5: Motion

1. the rate at which an object changes velocity; change in velocity divided by time
2. when two or more moving objects run into one another
3. the motion of an object; mass times velocity
4. the rate at which an object moves; distance divided by time
5. a quantity that includes magnitude and direction
6. the speed of an object in a given direction
7. B.
8. B.

Lesson 6: Forces

1. B.
2. B.
3. The person would weigh more on Earth because the planet has more mass and, therefore, a higher force of gravity.
4. The force of gravity is pulling on the book, but the table is pushing back with an equal and opposite force. Newton's Law: an object at rest will remain at rest until a force is acted upon it.
5. The rope will move to the right, because Javier is pulling with a greater force than Andy.
6. The 500-kg truck will hit the wall with more force because it has a greater mass, and force is equal to acceleration times mass.
7. reaction
8. motion
9. greater
10. greater

Lesson 7: Work

1. joule
2. complex machine
3. Energy
4. lever — brakes, wheel and axle — wheels, pulley — gears, screw — hold frame together
5. When Carrie pushes on the log alone, no work is being done since the log does not move. When Nina and Carrie push the log together, no work is being done since the log does not move. When Nina, Carrie, and Pia all push on the log, work is being done on the log since the log is displaced.
6. C.

Lesson 8: Structure of Matter

1. Pure substances are made entirely from identical atoms or molecules; you can't change the ratio of the compounds without changing the identity of the substance. Mixtures can have different ratios of their components. They can also be separated into their components.
2. Homogeneous mixtures look uniform throughout, while the different components in heterogeneous mixtures can be seen.
3. C.
4. B.
5. an element
6. heterogeneous
7. mixture
8. filtration

Lesson 9: Properties of Matter

1. A physical property can be observed without changing the identity of the substance. A chemical property can be observed only when the substance is part of a chemical reaction, which changes the identity of the substance.
2. You can observe and measure how shiny a metal is without changing its identity.
3. It is a physical change because, during a phase change, a substance changes from one state to another but does not change its identity.
4. C.
5. B.
6. B.

Lesson 10: Chemical Reactions

1. Al and Cu are solid; $CuCl_2$ and $AlCl_3$ are aqueous solutions.
2. Aqueous $AlCl_3$ and solid Cu are the products, and solid Al and aqueous $CuCl_2$ are the reactants.
3. Each side has a total of 11 atoms; 2 Al, 3 Cu, and 6 Cl.
4. Solid aluminum reacts with aqueous copper chloride to form aqueous aluminum chloride and solid copper.
5. The chemical identity of the reactants changes, but the total mass stays the same.
6. 5 on the left and 1 on the right

7. 12 on the left and 2 on the right
8. 2 on the left and 3 on the right
9. 19 on the left and 6 on the right
10. D.
11. $AgNO_3$
12. atoms
13. 55.85
14. 40.08

Lesson 11: Solutions

1. A saturated solution has exactly the amount of solute that can be dissolved at that particular temperature. An unsaturated solution has less solute than the amount that would make it saturated. A supersaturated solution temporarily has more solute than can be dissolved at that temperature.
2. An acid forms hydrogen ions when added to water; acids lower the pH of water. A base forms hydroxide ions when added to water; bases raise the pH of water.
3. B.
4. C.
5. 30; 50°C
6. less
7. an acid
8. a base

Unit Practice Test

1. A.
2. more
3. D.
4. hydroelectric
5. nuclear
6. conduction
7. exothermic
8. A.
9. C.
10. radio waves, x-rays, gamma rays
11. Blue car; it traveled the 800 meters in the shortest amount of time.
12. A.
13. A.
14. Work was being done on the bundle of lumber only after five people were trying to lift it. This is because it is the only time the bundle was displaced.
15. D.
16. B.
17. B.
18. C.
19. B.
20. Both the toy car and the toy truck have work being done on them since both are displaced in this scenario.
21. The toy car will move with a greater acceleration because it has less mass.
22. D.
23. filtration
24. C.
25. pure substance
26. B.
27. - ; 0
28. liquid
29. 0.89
30. D.
31. physical
32. physical
33. chemical
34. atoms
35. exactly the same
36. D.
37. supersaturated
38. higher
39. concentrated
40. a base
41. stronger

Unit 4: Earth Science

Lesson 1: Matter Cycles

1. a natural cycle that moves carbon through the atmosphere, soil, plants, and animals
2. a process in which bacteria change nitrates back into nitrogen gas
3. natural resources such as coal and oil
4. materials found on Earth that are useful to humans
5. a natural cycle that moves nitrogen through the atmosphere, soil, plants, and animals
6. C.
7. B.
8. D.

Lesson 2: Characteristics of the Atmosphere and Oceans

1. the apparent curving of wind and water due to the rotation of Earth
2. large-scale movements of water across Earth's surface
3. a condition of the atmosphere and ocean system caused by unusually high ocean temperatures
4. a moving spiral of water around the oceans caused by the Coriolis effect
5. B.
6. B.
7. C.

Lesson 3: Interior Structure of Earth

1. transform
2. crust
3. D.
4. A.
5. D.

Lesson 4: Weathering and Natural Hazards

1. a process that changes the composition of rocks by means of chemical reactions
2. a sudden vibration in the earth in which plates slip past one another
3. a large storm with winds greater than 75 mph
4. a process that breaks down rocks but does not change the composition of the rocks
5. the breaking down and wearing away of rocks
6. A.
7. D.
8. B.

Lesson 5: Age of the Earth

1. the naturally preserved remains or traces of animal or plant life from the past
2. natural formations that occur on Earth
3. methods of using the decay rate of materials in rocks and soil to learn how long they have been there
4. carbon-14
5. body
6. D.
7. C.
8. canyon, continent, mountain range, coast
9. carbon-14, uranium-235
10. wind, water

Lesson 6: Structures in the Universe

1. a pattern of stars in the sky that humans have observed and named
2. a collection of dust, gas, and billions of stars held together by gravity
3. one star and all the objects that orbit it
4. a bright sphere of gas held together by its own gravity
5. B.
6. C.
7. expanding
8. gravity
9. Neptune

Lesson 7: Structures in Our Solar System

1. A lunar eclipse occurs when Earth moves between the sun and the moon and casts a shadow on the moon.
2. a solid object that orbits a planet
3. a large, round body of matter that orbits a star
4. A solar eclipse occurs when the moon moves between Earth and the sun and blocks the view of the sun from Earth.
5. the periodic rising and falling of water levels on Earth caused primarily by the moon's gravity
6. B.
7. C.
8. A.

Unit Practice Test

1. B.
2. Possible answers: Trees would not be able to remove carbon dioxide from the atmosphere. The burning of the trees would add more carbon dioxide to the atmosphere.
3. A.
4. A.
5. Possible answer: As the water in the polar regions warms, the sinking of cold water can be slowed or stopped. This would stop the water from going back toward the equator, resulting in warming temperatures around the world. This, in turn, would further stop the formation and sinking of cold water at the poles.
6. C.
7. D.
8. Possible answer: The carbon-14 method is used for younger organic matter. Carbon-14 is a weakly radioactive form of carbon that is found in small amounts in plants and animals. The amount of carbon-14 is constant in living organisms. When a living organism dies, the amount of carbon-14 inside it begins to decrease. Scientists can use this information to estimate how long ago the organism died.
9. physical weathering: ice wedging
   chemical weathering: acid rain
   natural hazard: earthquake
10. Mercury, the moon, Neptune
11. D.
12. C.
13. zero
14. solar

GLOSSARY

acceleration — the rate at which an object changes velocity; change in velocity divided by time

acid — substance formed from hydrogen ions in water

adaptation — a trait that is found in most members of a species

allele — a different form of a gene

amino acid — a molecule that is the building block of proteins

antibody — a substance that the immune system produces to destroy pathogens

antigen — a protein on the surface of a pathogen

atom — the smallest particle into which an element can be broken down and still have the properties of the element

atomic mass unit (amu) — the average mass for one atom of an element

bacteria — a one-celled organism that has DNA

balanced equation — a chemical equation in which the same number of each type of atom appears on both sides

base — substance formed from hydroxide ions in water

bias — anything that sways an experiment's results in a way that makes them inaccurate

calorie — a measurement of energy

carbon cycle — a process that keeps the carbon on earth in balance

carrying capacity — the maximum population an ecosystem can support without losing resources

cell — the basic unit of life

cell cycle — the process of division of a cell into new identical cells

chart — a visual representation of data

chemical equation — a representation of what occurs during a reaction

chemical change — a change in which the chemical identity of the substance is changed

chemical property — a property that can be observed only when a substance takes part in a chemical reaction

chemical reaction — a process in which the atoms of one or more substances rearrange and change how they are connected, producing at least one new substance

chemical weathering — process that changes the composition of the rocks and occurs through chemical reactions

chromosome — a structure in a cell that contains DNA

codon — a combination of three bases

coefficient — a number added to the reactants and products to balance a chemical equation

collision — when two or more moving objects run into one another

commensalism — a relationship that benefits one organism, while the other is neither harmed nor helped

common ancestor — an individual in a species history to which all individuals in that species can be traced

complex machine — two or more simple machines put together

compound — a pure substance containing a specific ratio of two or more elements chemically bonded

concentrated — when a solution contains a large amount of solute

conclusion — a decision reached by using data and information

conduction — transfer of heat through direct contact

constellation — a pattern of stars in the sky that humans have observed and named

consumer — an organism that must eat other organisms to obtain energy

control variable — a factor that is kept the same or constant during an experiment

convection — the movement of heat through a liquid or gas

Coriolis effect — the apparent change in motion of wind and water on Earth's surface due to rotation

crust — the uppermost layer of Earth

current — a large movement of water

data — information gathered during an experiment or investigation

decomposer — an organism that feed off the dead bodies of a once-living organism

denitrification — a process that converts nitrates back to nitrogen gas

density — the ratio between the volume of a sample of a substance takes up and the mass of that sample

dependent variable — a factor that is changed in response to the independent variable during an experiment

diagram — a picture or illustration that shows information

digestion — the process of breaking food down into components the body can use

dilute — a solution with a small amount of solute

distillation — a process for separating liquids that have different boiling points

DNA — a double-stranded substance that contains the genetic "blueprint" of an individual

dominant — an allele for which the phenotype will always be present

double blind — an investigation in which neither the researchers nor the patients know who is receiving placebos

earthquake — a sudden vibration in the earth in which plates slip past one another

GLOSSARY

ecosystem — a community of living organisms and nonliving things that work together to make a balanced system

El Niño — an uncommon condition caused by unusually warm temperatures in the tropical Pacific Ocean

electromagnetic spectrum — the entire spectrum of radiation, ranging from radio waves to gamma rays

electrons — particles with a negative charge

element — pure substance that consists of smaller particles

endothermic reaction — a reaction in which more energy is absorbed than released

energy — the ability to do work

energy pyramid — a model of the energy flow through an ecosystem

enzyme — a large molecule that speeds up a chemical reaction

epigenome — chemical tag covering the genome

error — mistake in an investigation, including inaccuracy of measuring instruments and sampling

evaporation — the change of a substance from a liquid to a gas

evidence — information used to show whether a conclusion is valid

exothermic reaction — a reaction in which more energy is released than absorbed

fermentation — the process of producing energy when there is not enough oxygen to break down sugar

filter — a tool used to separate a mixture

filtration — a process for separating a mixture

fission — the process of splitting an atom

food chain — a linear path within a food web

food web — a model of the feeding connections among species

force — a push or a pull that results from an object's interaction with another object

fossil — naturally preserved remains or traces of animal or plant life

fossil fuel — fuel formed from the remains of organisms

frequency — how often the particles vibrate as a wave passes through a medium

galaxy — a collection of dust, gas, and stars held together by gravity

gene — a segment of a chromosome that has codes for a particular trait or function

generator — a device that converts mechanical energy into electrical energy

genome — an organism's entire set of DNA

genotype — the genetic makeup of an organism

gland — an organ that secretes chemicals to be used by the body

graph — a visual representation of data

gravitational potential energy — the energy an object possess because of its gravity

gravity — the force that pulls downward toward an object's center

greenhouse gas — a gas that increases the greenhouse effect by absorbing infrared radiation

gyre — a current that moves in spirals

hardness — the measurement of force needed to break a substance

heat — a type of energy

heterogeneous — a non-uniform mixture, such as oil and vinegar

homeostasis — the tendency of a system to achieve a relatively stable equilibrium

homogeneous — a uniform mixture, such as salt and water; also called a solution

hormone — a chemical produced by the body that serves a regulatory role

host — an organism on which a parasite lives

hurricane — a large storm with very strong winds more than 75 mph

hydroelectric power — electrical power produced from the mechanical energy found in moving water

hypothesis — a prediction or an educated guess that answers a question

immune system — the group of organs that protects the body from infection

immunization — a process in which the body receives a vaccine

inclined plane — a simple machine consisting of a flat, slanted surface

independent variable — a factor that is changed by a scientist during an experiment

inertia — the tendency of an object to stay at rest or in motion unless acted upon by an outside force

infection — the presence of a pathogen in the body

infrared radiation — energy in the electromagnetic radiation spectrum with wavelengths longer than those of visible light and shorter than those of radio waves

inner core — the solid inner layer of Earth's core

invasive species — a plant or animal that is not native to an area and has a negative effect on that area

joule — unit of work

kinetic energy — the energy an object has because of its motion

landform — natural formations that occur on Earth

Law of Conservation of Energy — law that states that energy can be transformed but cannot be created or destroyed

Law of Conservation of Matter — law that states that matter cannot be created or destroyed

lever — a simple machine consisting of a rigid arm that pivots on a fulcrum

limiting factors — a factor that limits the size of a population

limiting reactant — the reactant that is used up first in a chemical reaction

lunar eclipse — an event in which Earth moves between the sun and the moon and casts a shadow on the moon

machine — tools that make doing work easier

mantle — semi-solid layer of rock found under the crust

mechanical energy — the sum of an object's kinetic and potential energy

meiosis — a form of cell division by which a single cell divides into four daughter cells

metabolism — the sum of all chemical reactions that take place in an organism to maintain life

minerals — inorganic substances needed in trace amounts for growth and health

mitosis — a form of cell division by which a single cell divides into two identical cells

mixture — a substance made of two types of substances together

model — a representation of facts or information

molecule — a substance made of two atoms bonded together

momentum — the motion of an object; mass times velocity

moon — a solid, spherical object that orbits a planet

mutation — a change in DNA

mutualism — a relationship between two organisms in which both organisms benefit

natural resources — materials formed in nature that are useful to humans

natural selection — the process in which the individuals with beneficial traits survive

negative feedback — a reaction to change in a way that brings it back to its original state

nerves — a bundle of nerves that transmits information in the body

neutral — when a solution has a pH of 7

neutron — a particle with no charge

Newton's First Law of Motion — states that an object will tend to stay at rest or in motion unless acted upon by an outside force; also known as inertia

Newton's Second Law of Motion — states that the greater the mass of an object, the greater the force needed to move that object

Newton's Third Law of Motion — states that for every force applied to an object, an equal force acts in the opposite direction

nitrogen cycle — a process that moves nitrogen through Earth's systems

nonrenewable resource — a resource that cannot be readily replaced by natural means

nuclear energy — the energy released during nuclear fission or fusion reactions, generally used to produce electricity

nucleus — the center of an atom

nutrient — a substance that provides nourishment to a living thing

organelle — smaller structure found within a cell

organ — a substance in the body that is made of different tissues that work together to perform a specific function

outer core — the liquid outer layer of Earth's core

parasitism — a relationship that helps one organism while causing harm to the other

pathogen — an organism that can cause a disease in a living thing

periodic table — graphic representation of all known elements

pH — scale used to indicate how acidic or basic a solution is

phase — matter in the form of a solid, liquid, or gas

phase change — a physical change to the shape of a substance

phenotype — the form of a trait that is visible in an organism

photosynthesis — a process in plants that converts light energy to chemical energy

physical change — a change to a substance that does not change its chemical identity

physical properties — properties that can be observed without changing the chemical identity of the substance

physical weathering — a process that breaks down rocks but does not change the composition of the rock

planet — a large, spherical body that orbits a star

potential energy — the energy an object possesses due to its position relative to other objects

power — the rate at which work is done

precipitate — when a solute begins to come out of solution

prediction — a guess based on facts

GLOSSARY

producer — an organism that capture the sun's energy and make their own food

products — the symbols on the right side of a chemical equation

protein — a long molecule used to build body tissues and send signals

proton — a particle with a positive charge

pulley — a simple machine consisting of a rope that fits into a wheel that can be moved

Punnett Square — a diagram that shows the possible combinations of genotypes two parent organisms can produce

pure substance — a substance made of identical atoms or molecules

radiation — the movement of heat without particles

radiometrics — methods of using the decay rate of materials in rocks and soil to learn how long they have been there

reactants — the symbols on the left side of a chemical equation

recessive — an allele for which the phenotype will be present only if the genotype is solely that allele

renewable resource — a resource that can be readily replaced by natural means

respiration — the chemical processes that allow an organism to gain energy from nutrients

Ring of Fire — an area of active tectonic activity along the rim of the Pacific plate

RNA — a single-stranded substance that carries cellular information

saturated — the point at which a solute will no longer dissolve in a solvent

screw — a simple machine consisting of an inclined plane wrapped around a cylinder

solar eclipse — an event in which the moon moves between Earth and the sun, blocking the view of the sun

solar system — a star and all the objects that orbit it

solubility — a physical property indicating how much of a substance can be dissolved

solute — the substance being dissolved

solvent — what a substance is being dissolved in

speciation — the process through which a group becomes two or more different species

speed — the rate at which an object moves; distance divided by time

star — bright spheres of gas held together by their own gravity

strong acid — an acid that forms a high number of hydrogen ions in a water solution

strong base — a base that raises the pH of a solution very high

subduction — the movement of one tectonic plate under another

supersaturated — a solution that has exceeded the point of saturation

symbiosis — any relationship between two organisms

system — a combination of organs and tissues that work together to perform a specific function

table — use of columns and rows to show data

tectonic plates — pieces of lithosphere that hold the continents and oceans

temperature — a measure of how hot something is

thermal energy — heat energy

tide — the periodic rising and falling of water levels on Earth

tissue — substance made of similar cells that work together to perform a specific activity

toxin — poison

transcription — the process of making an RNA copy of a DNA segment

translation — the process in which cells make proteins

transmit — to pass something on to another

trophic structure — the feeding relationship between organisms in an ecosystem

turbine — a machine that converts kinetic energy into mechanical energy

unsaturated — when a solvent can dissolve more solute

vaccine — a dead or inactive pathogen

variation — differences within individuals of a species

vector quantity — quantity that includes the magnitude and direction

velocity — speed of an object in a given direction

virus — a pathogen that can only reproduce inside the cell of a living thing

vitamin — organic substances needed in trace amounts for growth and health

watt — unit of power

wavelength — the distance between two successive points in a wave

weak acid — an acid that forms a small number of hydrogen ions for the same amount of solute

weak base — a base that raises the pH of a solution above 7

weathering — breaking down and wearing a way of rocks

wedge — a simple machine consisting of two inclined planes pushed back to back

wheel and axle — a simple machine consisting of a wheel and an axle

work — an action that causes displacement of an object